ROUTLEDGE LIBRARY EDITIONS:
WAR AND SECURITY IN THE MIDDLE EAST

Volume 4

THE GULF WAR AND THE ENVIRONMENT

THE GULF WAR AND THE ENVIRONMENT

Edited by
FAROUK EL-BAZ AND R. M. MAKHARITA

Routledge
Taylor & Francis Group
LONDON AND NEW YORK

First published in 1994 by Gordon and Breach Science Publishers S.A.

This edition first published in 2017
by Routledge
2 Park Square, Milton Park, Abingdon, Oxon OX14 4RN

and by Routledge
711 Third Avenue, New York, NY 10017

Routledge is an imprint of the Taylor & Francis Group, an informa business

British Library Cataloguing in Publication Data
A catalogue record for this book is available from the British Library

ISBN: 978-1-138-19428-1 (Set)
ISBN: 978-1-315-54183-9 (Set) (ebk)
ISBN: 978-1-138-64756-5 (Volume 4) (hbk)
ISBN: 978-1-138-65730-4 (Volume 4) (pbk)
ISBN: 978-1-315-62141-8 (Volume 4) (ebk)

THE GULF WAR AND THE ENVIRONMENT

Edited by

Farouk El-Baz

and

R. M. Makharita

GORDON AND BREACH SCIENCE PUBLISHERS

USA • Switzerland • Australia • Belgium • France • Germany • Great Britain
India • Japan • Malaysia • Netherlands • Russia • Singapore

Gordon and Breach Science Publishers S.A.
World Trade Center
Case postale 531
1000 Lausanne 30 Grey
Switzerland

Library of Congress Cataloging-in-Publication Data

The Gulf War and the environment / edited by Farouk El-Baz and R.M. Makharita
p. cm.
ISBN 2–88124–649–4. — ISBN 2–88449–100–7 (pbk.)
1. Persian Gulf War, 1991—Environmental aspects—Persian Gulf Region. I. El-Baz, Farouk. II. Makharita, R. M.
DS79.744.E58G85 1994
956.704'421—dc20 93–49407
CIP

CONTENTS

PREFACE

The Gulf War of 1991 was perhaps the first explicitly environmental war of the modern era. Spilling of an estimated total of 8 million barrels of oil and wanton explosions of 732 oil wells created a disaster of regional proportions. Much of the damage to the environment was vividly displayed by the televised media worldwide. People everywhere watched as birds perished in the thick oil that floated on the Gulf water, and as humans attempted to escape the suffocating plumes of the burning oil wells of Kuwait.

Immediately following the cessation of hostilities between Iraq and coalition forces from thirty countries, scientists from Kuwait and other countries in the Gulf region began to study the effects of the war on their land and coastal waters. Because of the enormous magnitude of the problem, they were assisted by teams of investigators from other countries. One such mission was conducted in April 1991 on behalf of the Third World Academy of Sciences in Trieste, Italy, to evaluate the environmental effects of the war on the six countries of the Gulf Cooperation Council—Kuwait, Saudi Arabia, Bahrain, Qatar, the United Arab Emirates and Oman. During this fact-finding mission it was decided to convene a scientific conference within one year to evaluate the environmental effects of the Gulf War on the region.

Local, regional and international organizations cooperated by holding a "Workshop on the Environmental Dimensions of the Gulf: Policy and Institutional Perspectives." The workshop was hosted in April 1992 by the United Arab Emirates University at Al Ain, U.A.E., with financial support from The World Bank, Washington, D.C. The senior editor of this volume acted as organizer of the scientific sessions, which brought experts from the Gulf region together with others from the international scientific community. The papers in this volume were presented at the workshop.

This book serves as a source of scientifically valid information collected soon after the end of the Gulf War. It will be of value to students, teachers, and organizations interested in environmental issues; environmental engineers and economists interested in damage assessment; public health experts and decision makers concerned with environmental quality; and legislators of environmental policies. In addition, it serves as documentation of the effects of war on arid environments due to disturbances to the natural protective layer on desert surfaces.

The editors wish to thank the forty organizations that participated in the workshop. Thanks are also due to His Highness Shaikh Nahyan Bin Mubarak Al-Nahyan, minister of education and scientific research and chancellor of the United Arab Emirates University, for hosting the workshop. We are particularly indebted to Paula Leonardo and Caroline de Bourbon of the Center for Remote Sensing at Boston University for typing the manuscript and assistance in its editing, respectively.

Farouk El-Baz

R. M. Makharita

DETECTION BY SATELLITE IMAGES OF ENVIRONMENTAL CHANGE DUE TO THE GULF WAR

FAROUK EL-BAZ, A. ABUELGASIM, M. KOCH, M. PAX-LENNEY, E. LAMBIN, A. AL-DOASARI, P. MARR, S. RYHERD, and R. MORENCY

Center for Remote Sensing
Boston University, Boston MA 02215

The invasion of Kuwait on 2 August 1990, the subsequent occupation by Iraqi military forces, and the war of liberation that followed, have affected the environment of the northern part of the Arabian Gulf region. Environmental impacts on the atmosphere from the burning of Kuwait's oil wells, on the Gulf water from major oil spills, and on the land surface from disturbances of the desert pavement are clearly depicted on satellite images. Photographs by Meteosat, NOAA 10 and 11 satellites and the Space Shuttle astronauts and particularly images by the Landsat Thematic Mapper and SPOT show the utility of numerous images from various space platforms. Change detection techniques are very useful in recognizing environmental change through the comparison of pre-war and post-war images. Furthermore, Geographic Information System (GIS) methodologies allow production of thematic maps to distinguish change and interpret its causes.

I. INTRODUCTION

Iraq invaded Kuwait on 2 August 1990. During the eight months that followed, war preparations and the military conflict that ensued resulted in much degradation to the environment of the region. Environmental impacts resulted in much degradation to the environment of the region including the burning of Kuwait's oil wells, spilling of oil in the Arabian (Persian) Gulf water, and disturbing the desert surface by military activities. The latter was only recently recognized as a war-related environmental impact based on observations in the Sinai Desert of Egypt (Holden, 1991).

Severe environmental damage and disruption of operations in Kuwait's oil fields (Figure 1) resulted from the explosion of 732 wells by Iraqi troops before they left Kuwait. Efforts by teams of fire fighters succeeded in capping the last well fire on 8 November 1991 (Sipa, 1992, p. 162). This ended the dark plume of oil droplets, soot and smoke that emanated from the well fires, and ceased the "oil rain" that blackened vast stretches of the desert and polluted coastal areas of the Gulf. The predominance of oil droplets in the "smoke plume" from the well fires was first noted by El-Baz (1992a and b) after a visit to the region on 23 April–3 May 1991 (Table 1).

As the well fires were put out, much oil seeped from the well-heads and formed stagnant pools. Furthermore, some of the wells exploded by Iraqi troops did not catch fire and continued to spew oil onto the land creating oil patches, streams and vast lakes (El-Baz, 1992c). These lakes hamper operations in the oil fields and constitute a health hazard to humans, animals and plants. Furthermore, roadbeds that had to be built to reach the oil

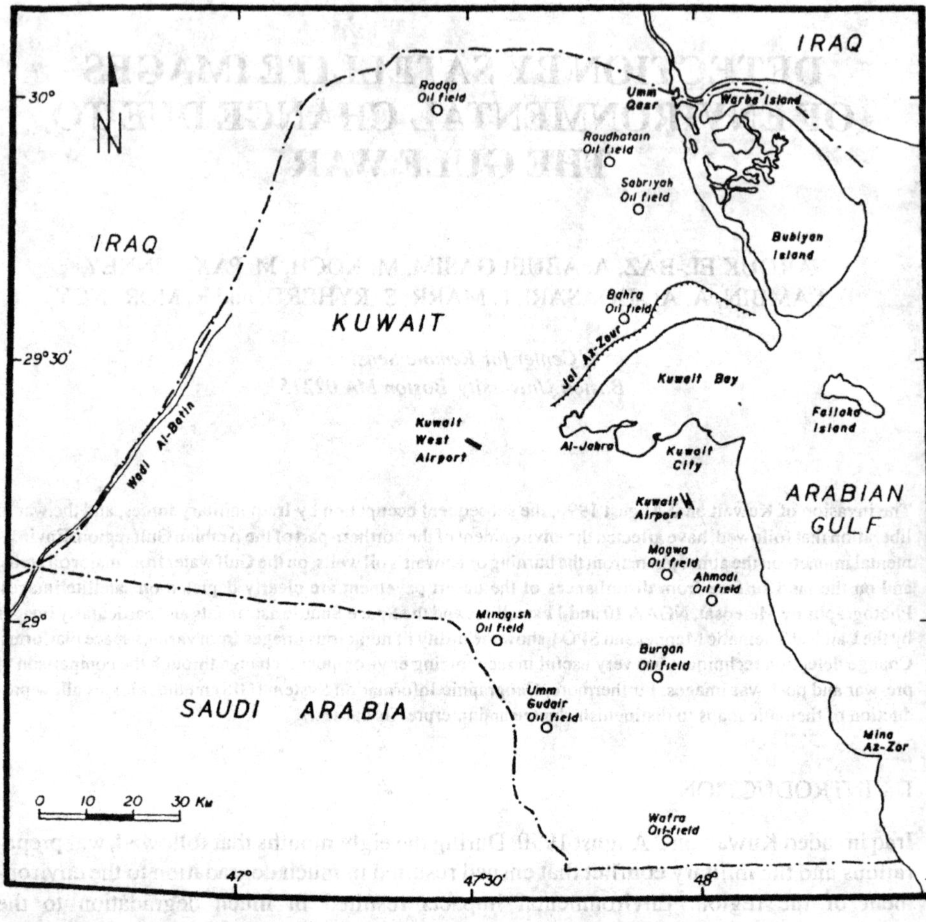

Figure 1. Map of Kuwait showing its major geographic features and the location of ten oil fields.

wells during firefighting operations cause further hazards as they are affected by wind erosion and sand deposition. All of these features are clearly depicted in satellite images of Kuwait and neighboring regions.

II. SATELLITE IMAGES

As the introduction of aircraft, early in this century, has revolutionized the map-making process by providing photogrammetric quality aerial photographs, space age technology has made another giant leap by providing similar photographs from space. During the past 25 years, much has been learned about photographing the Earth from space. The American Gemini, Apollo, Skylab, Apollo-Soyuz, and Space Shuttle missions returned various photographs of the Earth utilizing numerous instruments (El-Baz, 1984). In addition, the Landsat program introduced in 1972 digital imaging from space, where image data were transmitted to ground receiving stations. The technology of these systems provides an ad-

Table 1. Chronology of the Gulf Conflict and Post-War Activities. (Modified from Williams et al., 1991)

2 August 1990	Iraq invades Kuwait
Late October-early November 1990	Iraqi forces change offensive posture to defensive fortifications; digging in the land surface of Kuwait begins in earnest
16 January 1991	Allied forces declare war and begin aerial bombardment of Iraq
19–23 January 1991	Iraq releases oil from Sea Island terminal into the Gulf
22 January 1991	Iraq initiates the burning of oil wells, starting with the Wafra field and moving northward
26 January 1991	Allied bombing ends oil flow from terminal; oil visible in gulf on NOAA AVHRR image
8 February 1991	Oil visible in Gulf on Landsat TM image
9 February 1991	Smoke plumes from oil well fires visible on SPOT image
15 February 1991	Smoke plumes visible on Landsat TM image
21 February 1991	Smoke plumes visible on NOAA AVHRR image
24 February 1991	Allied ground offensive begins; extensive smoke plumes visible in Landsat TM, Meteosat, and NOAA AVHRR image
2 March 1991	Smoke plumes visible on SPOT images
3 March 1991	Conclusion of ground campaign; smoke plumes, oil slick visible on Landsat TM image
8 March 1991	Smoke plumes visible on Meteosat image
16 March–4 April 1991	United Nations sends mission to assess loss of life in Kuwait; the mission also assessed damage to Kuwait's infrastructure (UN, 1991)
23–31 March 1991	UK Meteorological Office aircraft flights to study smoke plumes in the atmosphere of Kuwait and environs
5–11 April 1991	Space Shuttle mission STS-37 photographs plumes and oil slick
10 April 1991	Smoke plumes visible on NOAA AVHRR image
23 April–3 May 1991	Mission of Third World Academy of Sciences to the Gulf region
28 April–6 May 1991	Space Shuttle mission STS-39 photographs plumes and oil slick
16 May–15 June 1991	University of Washington aircraft flights to study smoke plumes
18 May 1991	Goethe University aircraft flights to study smoke plumes
19 May–4 June 1991	NCAR aircraft flights to study plumes
May–June 1992	NOAA sends the Mount Mitchell to study Gulf water and biota

vanced new tool for the acquisition of data that are necessary for monitoring environmental change on the surface of the Earth.

In the acquisition of data necessary for the study of environmental change, unmanned and manned spacecraft systems are planned to fly in high, medium, or low orbits. The highest orbits are left to the unmanned weather satellites, such as Meteosat. These are propelled to a height of 36,000 kilometers above the Earth. At this altitude, their motion is equivalent in speed to the rotation of the Earth about its own axis. Such satellites are termed geostationary, and remain above the same point on the Earth to acquire and transmit repetitive images as frequently as hourly. Due to their high altitude, the images they collect can cover most of one hemisphere of the Earth at low spatial resolution, which is ideal for studying global weather phenomena.

The intermediate orbits are those from 500 to 2000 kilometers above the Earth, the region where most unmanned imaging satellites are placed. For example, the polar-orbiting satellites of the National Oceanic and Atmospheric Administration (NOAA) fly at altitudes of 835 to 870 km; and the near-polar orbits of the American Landsat and the French *Systeme Pour l'Observation de la Terre* (SPOT) reach a maximum altitude of 920 km above the Earth. Images collected from those altitudes provide greater local detail than is possible from the high-altitude satellites, but the area covered by individual images is significantly reduced. One result of this is that these satellites provide less frequent coverage than weather satellites (16 days for Landsat, and 28 days for SPOT).

On the lower end, most manned missions are placed in orbits below 500 km, to a minimum of 150 kilometers above the Earth. For example, the Space Shuttle operational altitude is about 300 kilometers. From this altitude, images show great detail such as those of the Large Format Camera (LFC: Doyle, 1985; and El-Baz, 1985).

Images obtained from most of these satellite systems have been utilized in the present study of the disruption of the oil fields of Kuwait. For this reason, a description of the applicable systems is given below with emphasis on the data that pertain to the war's environmental impact; satellite images of the Gulf region obtained through June 1991 are listed by Williams et al. (1991).

1. *Meteosat*

Meteosat is a series of geostationary meteorological satellites launched by the European Space Agency (ESA); the first began operation in 1977. Two satellites are currently operating: Meteosat 3 is positioned at the Equator and 50 degrees west, above South America; and Meteosat 4 is positioned at the Equator and 0 degrees. Because the latter is positioned near the prime meridian, above West Africa, the satellite's view of the Gulf is oblique, since it is located far from the nadir (Figure 2).

The spin-stabilized, geostationary Meteosat spacecraft carry: (1) a visible-IR radiometer to provide high-quality day images and day and night temperature data; (2) a communication package to disseminate image data to user stations; and (3) a communication package to collect data from various Earth-based platforms. The radiometer acquires data in the visible, near-infrared, and thermal-infrared wavelengths. The two infrared images consist of 2500 lines of 2500 picture elements (pixels); the visible image has 5000 lines of 5000 pixels. The spatial resolution at the subsatellite point is 5 km for the infrared channels and 2.5 km for the visible band (Mason, 1987).

2. *NOAA Satellites*

The National Oceanic and Atmospheric Administration (NOAA) has two polar-orbiting Environmental Satellites. These are NOAA-10, which was launched in September 1986, and acquires data over the Gulf at approximately 7:30 am and 7:30 pm daily;and NOAA-11, which was launched in September 1988, and acquires data over the Gulf at approximately 1:40 am and 1:40 pm daily.

Primary sensors on the NOAA satellites include: an Advanced Very High Resolution Radiometer (AVHRR); TIROS Operational Vertical Sounder (TOVS); Earth Radiation

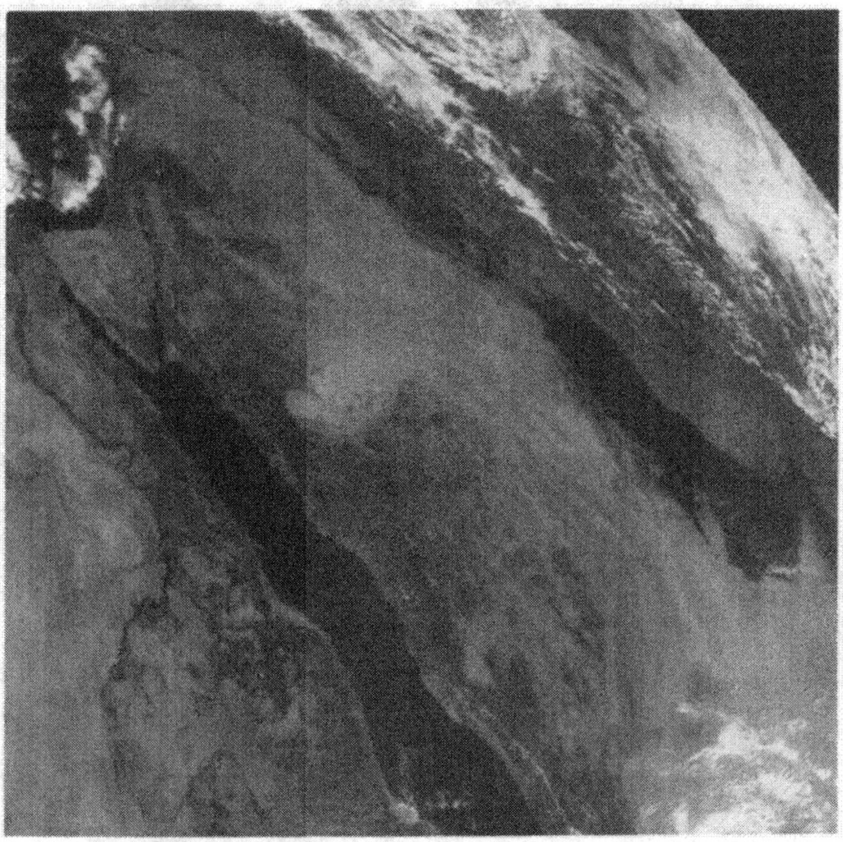

Figure 2. Meteosat image acquired on 10 February 1991, showing smoke plumes from Kuwait's oil well fires just west of the Arabian Gulf

Budget Experiment (ERBE); and a Solar Backscatter Ultraviolet Spectrometer (SBUV/2). Secondary sensors consist of a Space Environment Monitor (SEM), and a Data Collection System (DCS) in addition to a Search and Rescue system.

The primary environmental sensor on NOAA satellites is the AVHRR. It is a five-channel scanning radiometer that is sensitive in the visible, near-infrared, and thermal-infrared bands. Its data can be recorded in 1.1 km resolution (the basic resolution of the AVHRR instrument) or at 4 km resolution; the swath width is 2600 km. The stored high resolution (1.1km) imagery is known as Local Area Coverage (LAC). Owing to the large numbers of data bits, only 11 minutes of LAC can be accommodated on a single recorder. In contrast, 115 minutes of the lower resolution (4 km) imagery, called Global Area Coverage (GAC), can be stored on a recorder. The latter is enough to cover an entire 102-minute orbit of data (Rao et al., 1990).

One of the uses of the NOAA Environmental Satellite images of the Gulf region is their ability to display the oil slick on the Gulf water. As shown in Figure 3, the coastal zone of Kuwait was affected by such slicks.

Figure 3. Images obtained by the National Oceanic and Atmospheric Administration (NOAA) satellite 11, showing oil slick in the Gulf water off the coast of Saudi Arabia (courtesy of Robert M. Cary, NOAA). (COLOR PLATE I)

3. *Landsat*

Five satellites in the Landsat series have been launched since 1972 by the National Aeronautics and Space Administration (NASA). The satellites currently operating are Landsat 4 and Landsat 5, which are identical. The orbits are repetitive, circular, sun-synchronous, and near-polar at a nominal altitude of 705 km at the Equator. The satellites cross the Equator from north to south on a descending orbit at approximately 9:45 am on each pass. Each orbit takes nearly 99 minutes, and the spacecrafts complete just over 14 orbits each per day, covering the entire Earth (poles excepted) every 16 days. These satellites carry two instruments: a Multispectral Scanner System (MSS) and a Thematic Mapper (TM).

The Multispectral Scanner System (MSS) is an optical-mechanical line-scanning device that measures four spectral bands (green, red, and two wavelengths in the near-infrared region)with a fixed resolution of 82 m. The standard MSS scene is 185 km by 177 km. The coverage patterns result in 7.3% image sidelap at the Equator for Landsat 4 and 5 data. Image sidelap percentages increase proportionally as latitude increases.

The Thematic Mapper (TM) obtains images with a greater spectral, radiometric, and geometric sensitivity than the MSS. The TM is sensitive to seven spectral bands: three visible (blue, green, and red), near-infrared, two bands in the mid-infrared range, and one thermal-infrared band. The pixel resolution is 30 m for all bands expect the thermal-infrared band, which has a pixel resolution of about 120 m. The TM scene is approximately 185 km × 177 km (Hord, 1986).

Coverage of the Gulf Region

The total area of Kuwait is covered by four TM scenes (Figure 4). Each intersection of a path and row number represents a center of a scene. Landsat scenes of Kuwait that were acquired between January and October 1991 show plumes eminating from the well fires (Figures 5 and 6).

Also, Landsat images clearly depicted the oil spills in the water during the Gulf War. Prior to that war, the Gulf water and coastal zone had been altered by pollution from oil spills at oil terminals and from tankers. From this point of view, the Arabian Gulf is one of the most environmentally fragile bodies of water in the world. Although it covers 233,000 km and is nearly 1,000 km long, it is a semi-closed basin. The only connection to the open water of the Gulf of Oman and the Arabian Sea is the Strait of Hormuz, which is only 86 km wide.

The Gulf water is characterized by weak counter-clockwise currents, high temperature and high salinity as compared to the water of the Gulf of Oman. This forces the water that exits through the Strait of Hormuz to do so near the bottom, while replacement water floats on top. These characteristics make it difficult to cleanse the Gulf water naturally, because the cycle of water exchange takes several years.

The Gulf is very shallow, with an average depth of 35 m, and rarely exceeding 100 m. Coastal and nearshore oceanographic processes are complex, with significant temporal and spatial variations. The offshore water depth of the western part is generally less than 35 m, and the sea bottom is sandy. Depths, in nearshore waters up to 50 km offshore, rarely exceed 25 m and increase gradually with distance from shoreline (John et al., 1990).

For these physical and oceanographic reasons, oil spills in the Gulf water cause severe damage to its biological productivity (Al-Hassan, 1992). These spills also cause nearly

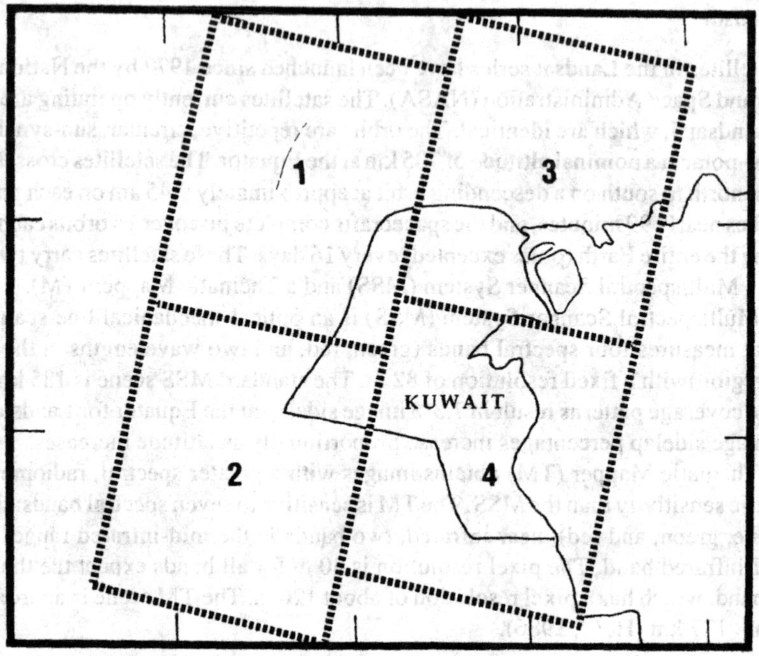

Figure 4. Schematic illustration of the area coverage of Kuwait by four Landsat Thematic Mapper images.

permanent damage to the coastline by polluting beaches with a thick deposit of tarry material, as occurred after the spill of 1983 from the Nowruz oil terminal that was damaged during the Iran/Iraq War (El-Baz, 1992a).

After initial sightings by military pilots during the early days of the Gulf War, the spilled oil was first observed in NOAA satellite images obtained on 26 January 1991 (Williams et al., 1991). Landsat images began to display spilled oil in images obtained on 8 February 1991. Following the counter-clockwise motion of the Gulf water stringers of oil began to appear along the east coast of Saudi Arabia south of Khafji.

It was immediately feared that the oil would endanger the 13 desalination plants on the western coast of the Gulf. Booms were extended and oil skimmers were emplaced to protect the intake of these plants. Over two million barrels of oil were pumped out from locations near Dhahran and Jubail in Saudi Arabia alone. However, the bulk of the oil was held northward at Abu Ali Island as shown in Figure 7.

4. *SPOT*

The SPOT system was designed by the *Centre National d'Etudes Spatiales* (CNES) and built by the French industry in association with partners in Belgium and Sweden. Like the American Landsat, it consists of remote sensing satellites and ground receiving stations. The imagining is accomplished by two High-Resolution Visible (HRV) instruments that operate in either a pancromatic (black-and-white) mode for observation over a broad spec-

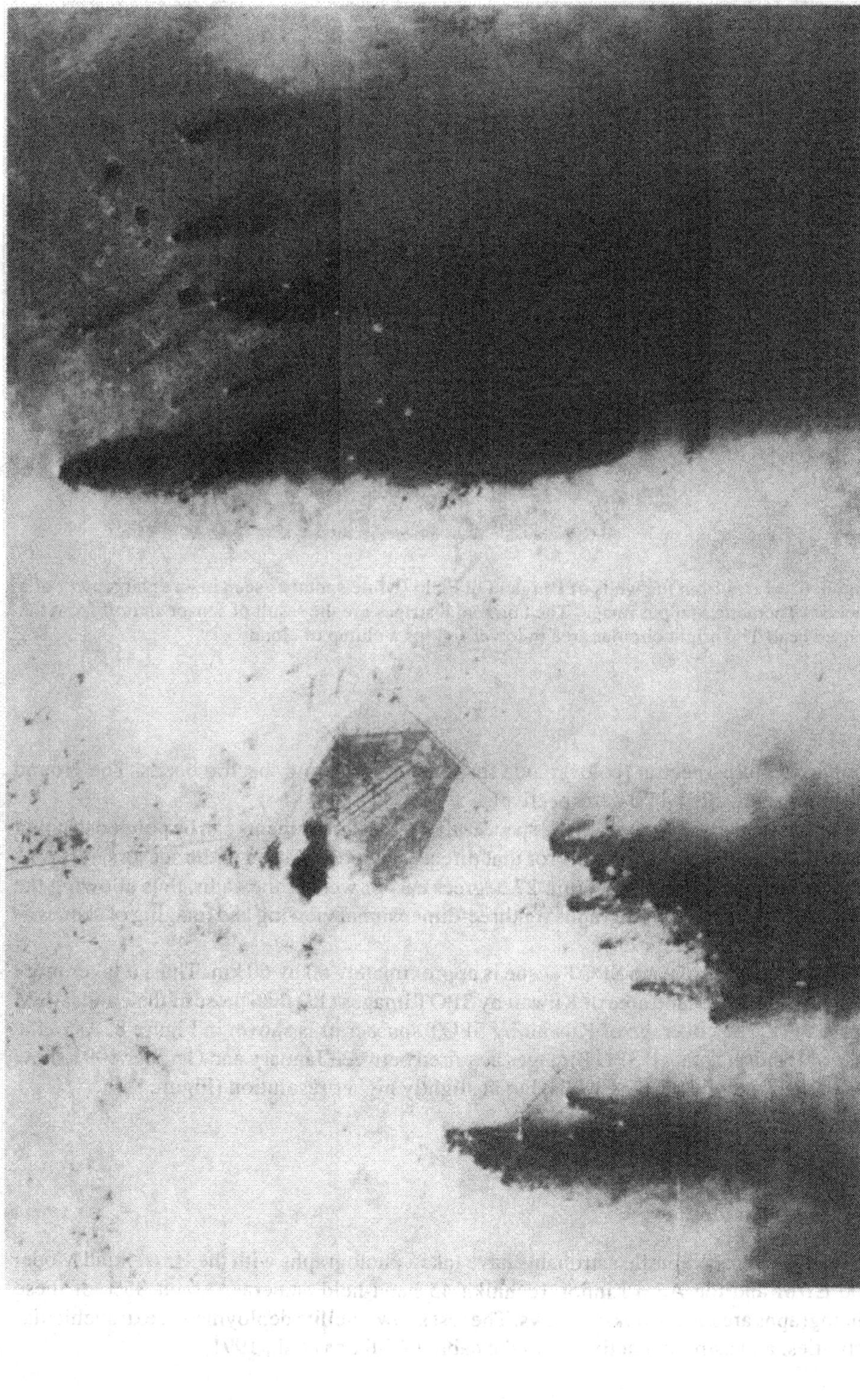

Figure 5. Smoke plumes from burning wells of the southern oil fields of Kuwait as depicted in a subscene of a Landsat Thematic Mapper image acquired on 15 February 1991.

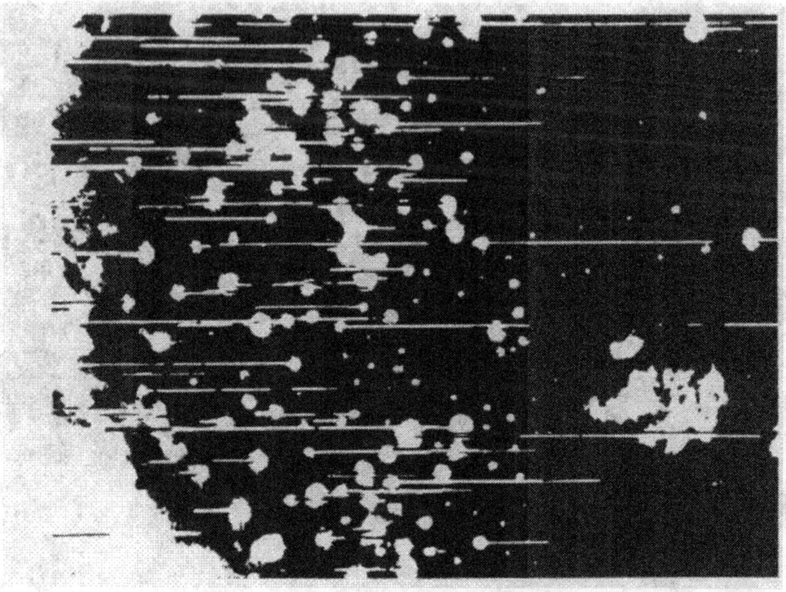

Figure 6. Fires from the wells of Burgan Oil Field (white spots) as seen in an enlargement of a Landsat Thematic Mapper image. The horizontal stripes are the result of sensor shutoff from the sensed heat. The bright circular area in lower right is a clump of clouds.

trum, or a multi-spectral (color) mode for sensing in narrow spectral bands. The ground resolutions are 10 and 20 m respectively.

For viewing directly beneath the spacecraft, the two instruments can be pointed to cover adjacent areas. By pointing a mirror that directs ground radiation to the sensors, it is possible to observe any region within 27 degrees east or west of the nadir, thus allowing the acquisition of stereo photographs for three-dimensional viewing and imaging of scenes as frequently as every four days.

The image area of each SPOT scene is approximately 60 by 60 km. Thus, it takes more scenes to cover the land area of Kuwait by SPOT images (13) than those of the Landsat TM images (4). The coverage of Kuwait by SPOT spacecraft is shown in Figure 8. As in the case of Landsat data, all SPOT images acquired between January and October 1991 show the plume from the burning wells, but at slightly higher resolution (Figure 9).

5. *Space Shuttle*

Since 1981, Space Shuttle astronauts have taken photographs with the Hasselblad Model 500 El/M and the Aero Linhof Technika 45 hand-held cameras. About 85% of these photographs are Earth-looking views. The rest show satellite deployments, extravehicular activities, and astronaut activities in the cabin (Williams et al., 1991).

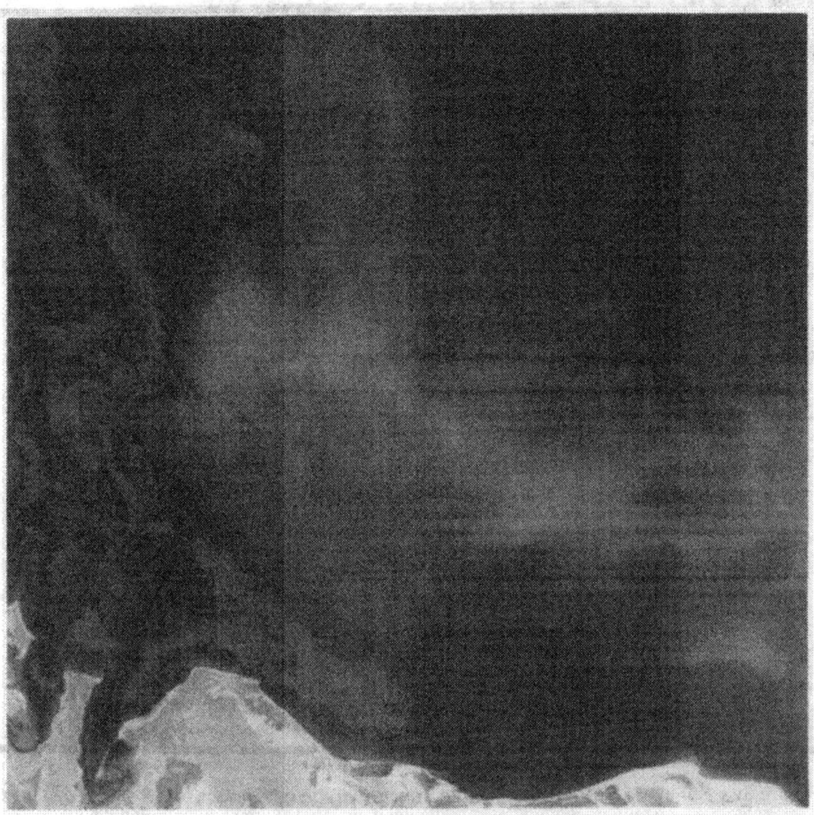

Figure 7. Oil slick in the Arabian Gulf waters off Manifa Bay, Saudi Arabia. Shades indicate variable concentration of oil and sheen as depicted on band 1 of a Landsat Thematic Mapper image acquired on 8 February 1991. (COLOR PLATE II)

Most of the photographs are in natural color, although a limited amount of black-and-white film has been used with polarizing filters. Beginning in 1983, a small amount of color infrared film was tested on some missions.

For most Shuttle missions, the orbital tracks cover the Earth between 28 N and 28 S latitude. Repeat coverage of an area is obtained by acquiring photography on several missions or by taking photographs from different viewing angles during a single mission. As a result of the Earth's rotation and the Shuttle's orbit duration, an area may be photographed at different sun angles during a single mission. Stereoscopic coverage is available for a number of areas (Lulla and Helfert, 1991).

Some of the photographs obtained by the Space Shuttle astronauts cover periods for which no Landsat or SPOT images were taken. These have proven very useful, particularly in displaying the effects of wind direction on the oil well fire plumes (Figure 10), as well as the spread of oil rain on the desert surface and oil lakes near damaged wellheads (Figure 11).

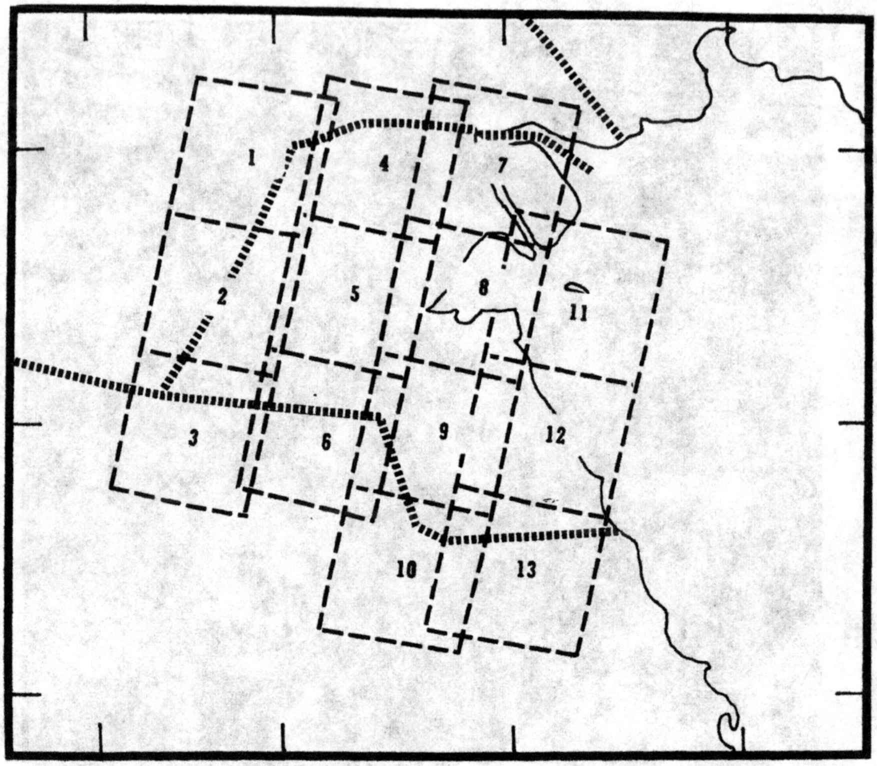

Figure 8. Schematic illustration of the area coverage of Kuwait by 13 images of SPOT.

III. CHANGE DETECTION

Change detection is the process of identifying differences in the state of a surface or a phenomenon by observing it at different times. The basic premise in using satellite images for change detection is that changes in radiance due to land cover change must be large with respect to radiance changes caused by atmospheric and other factors. Several procedures of surface change detection using digital data have been proposed. These methods include comparison of land cover classifications, multidate classification, image differencing/ratioing, principal component analysis, and change vector analysis (Singh, 1989).

Image differencing: In this method, spatially registered images of time t1 and t2 are subtracted, pixel by pixel, to produce a further image which represents the change between the two times. The technique yields a difference distribution of each band. In such a distribution, pixels showing radiance change are usually found in the tails of the distribution while pixels showing no radiance change tend to be grouped around the mean. The critical element of the image differencing method is deciding where to place the threshold boundaries between change and no change.

Image regression: In this technique, pixels from t1 are assumed to be a linear function of the time t2 pixels. The difference image can be found by subtracting the predicted

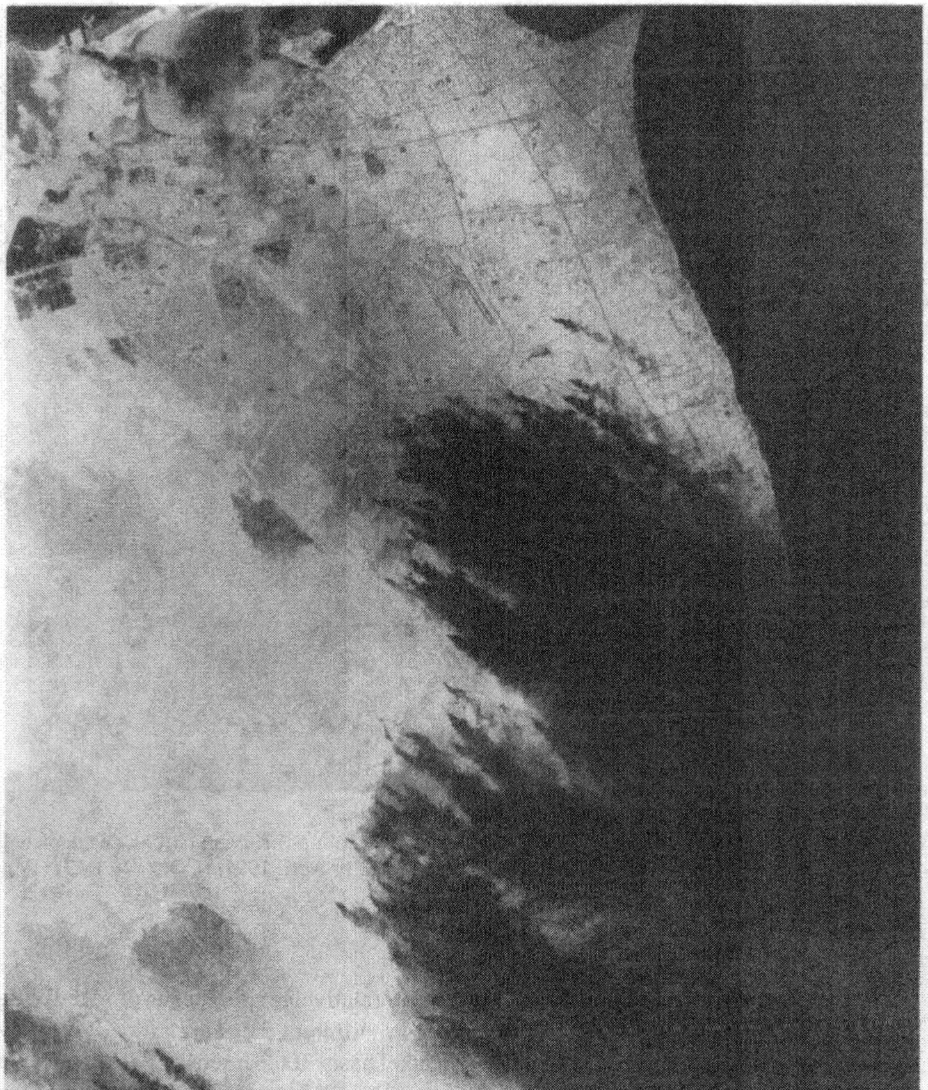

Figure 9. Plumes from oil well fires just south of the city of Kuwait as depicted on a SPOT image acquired on 2 March 1991. (COLOR PLATE III)

time t2 (obtained from the regression line) from the actual t2 image. A threshold technique is also followed to delineate areas of change from those of no change. The regression technique accounts for differences in the mean and variance between pixel values for different dates so that adverse effects from differences in atmospheric conditions and sun illuminations are reduced.

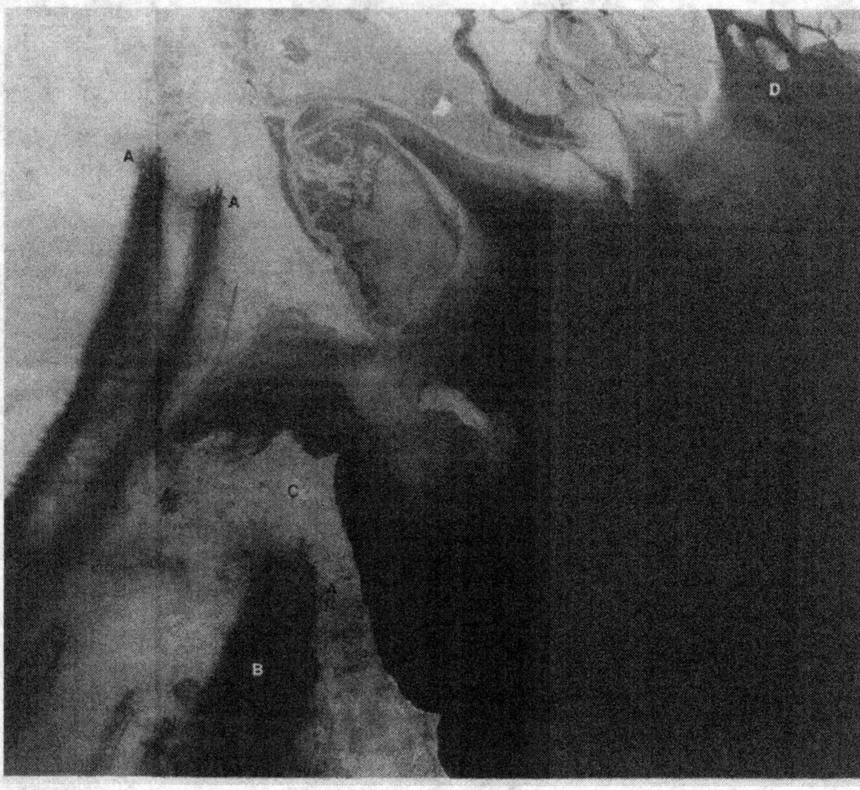

Figure 10. Plumes (B) of burning oil wells (A)north and south of Kuwait City (C) as depicted in a photograph acquired by Space Shuttle astronauts in early April 1991. (COLOR PLATE IV)

Image ratioing: Ratioing is considered to be a relatively rapid means of identifying areas of change. Here, two registered images from different dates are ratioed, band by band. The data are then compared on a pixel by pixel basis. If the intensity of the reflected energy is nearly the same in each image, then the ratio will be one; in areas of change, the ratio value would be significantly greater or less than one depending upon the nature of changes between the two dates. It is important to select appropriate threshold values in the lower and upper tails of the distribution representing change pixel values.

Principal component analysis: This technique is used to reduce the number of components to fewer principal components accounting for most of the variance in the original multispectral images. In multitemporal studies, the principal components for two or more dates are often compared as in image differencing or image regression. Two four-band Landsat scenes of the same area, which are recorded on different dates, can be superimposed and treated as a single eight band set. Principal component analysis of this data set should result in the gross differences associated with overall radiation and atmospheric changes.

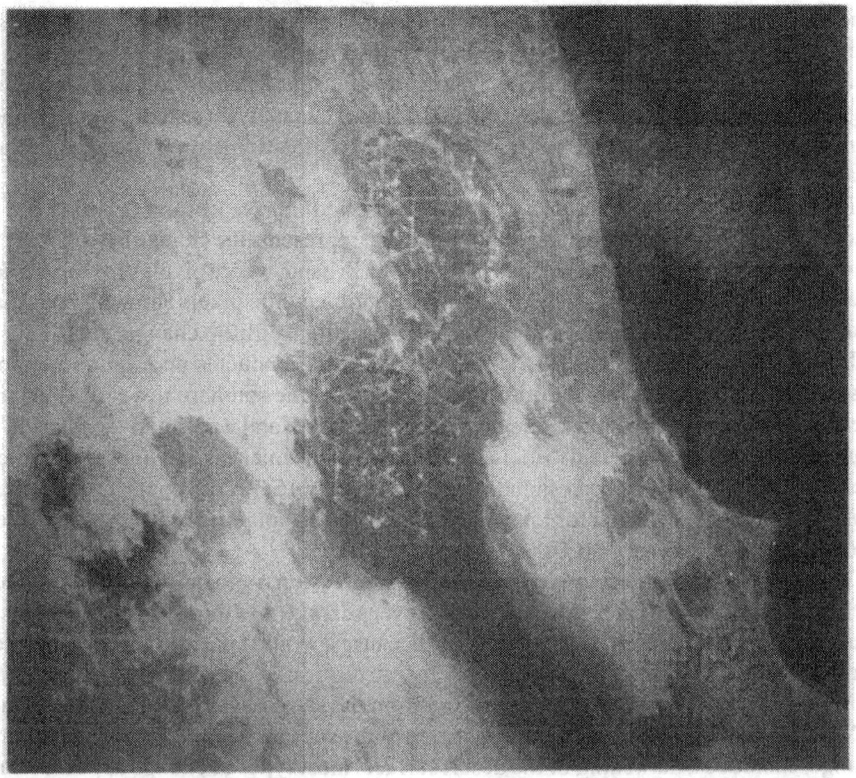

Figure 11. Photograph of the darkened surface of the desert in Kuwait due to oil and soot deposits from the fire plumes as depicted in a photograph acquired by Space Shuttle astronauts in December 1991, after the cessation of well fires.

Post-classification comparison: This method requires the comparison of independently produced classified images. By properly coding the classification results for times t1 and t2, a change map can be produced which shows a complete matrix of change. In addition, selective grouping of classification results allows the observation of any subset of changes which may be of interest. This method holds promise because the two data sets are classified separately, thereby minimizing problems of normalization for atmospheric and sensor differences.

Image registration is the most important and the most tedious task in digital image processing. Registration is the process of transforming/warping one set of data (usually called the uncorrected image) to register it with a second set of data (usually called the corrected or master image). The polynomial transform which describes how the uncorrected image data must be warped to fit over the master data is based on ground control points. After registration, the two image sets will have the same scale and rotation, and individual pixels will have exactly the same location in both scenes.

Landsat TM data obtained in January 1989 were utilized as the master images, and those obtained in June 1992 were considered uncorrected to test the utility of the technique in the study of changes in the oil fields of Kuwait. After image registration, the following steps are taken: (1) image regression to account for differences in the mean and variance between pixels of different dates due to changes in atmospheric conditions; (2) image ratioing to identify areas of change; (3) principal component analysis to reduce the number of components to principal changes; and (4) post-classification comparison to produce a change map.

In change detection techniques, spatially registered images of time t1 and t2 are subtracted, pixel by pixel, to produce a new image that represents the change between the two. The technique yields a difference distribution of each band, where pixels showing radiance change are usually found in the tail of the distribution, while pixels showing no radiance change tend to be grouped around the mean in situations of little change.

Using the software package EASI/PCI, a test was first conducted on a square area southeast of Kuwait International Airport. The site includes the southern town of Al-Ahmadi, the adjacent coastal strip, and a portion of the desert north and west of Al-Ahmadi. A portion of the Maqwa oil field is also included. The size of the image is 512 lines by 512 picture elements or pixels, which corresponds to approximately 15.36 km2 on the ground. Eleven control points were selected for image registration. The geometric registration resulted in a root mean square error of 0.12.

Three processes were examined for this site: regression, ratio, and subtraction. Due to extensive atmospheric haze, the 1992 image is very dark. It was thought that the regression analysis would be a preferred analytical tool because it would take into account the overall darkness of the 1992 image.

Band 4 of the 1989 image was subtracted from the same band of the 1992 image using PCI software. Figure 12 is the resulting image of the subtraction operation. Because most changes included a darkening of the desert surface, most types of change appear in darker tones while those areas or objects of little or no change are lighter.

A visual interpretation of the change image reveals oil lakes in the northwestern corner, exploded oil tanks in the central position, and darkened surface from oil rain in the western half of the image. The latter's trail follows the prevailing wind direction from the northwest to southeast. While this image is useful for interpretation, it does not provide quantitative assessments of change; each pixel of the change image must be interpreted and classified as change or no change according to its brightness value.

In order to determine boundaries or thresholds, a preliminary analysis of change at the pixel level was implemented. Twenty-seven sample pixels were selected within the image and covering the entire range of brightness values. The samples included areas of suspected change such as oil lakes, as well as areas thought to represent little or no change such as roads. The brightness values of each pixel were determined for the 1989, 1992, and the change images. A histogram of the change image was also produced.

Analysis revealed that all but one pixel of change fell to the left of the mean of the entire change image, i.e., less than 117; brightness values of Landsat pixels vary from 0 to 255. Therefore, the first threshold for change was set at 117 with values less than 117 indicating change. A second category of possible change was attempted based on the brightness values of pixels from a vegetated area. The assumption was that vegetation would not have been watered during the war, and would therefore, indicate change in the infrared from the 1989 to the 1992 image.

Healthy vegetation reflects strongly in the infrared. Lack of water would destroy some of the vegetation, resulting in a decrease in the infrared reflectance. Based on this informa-

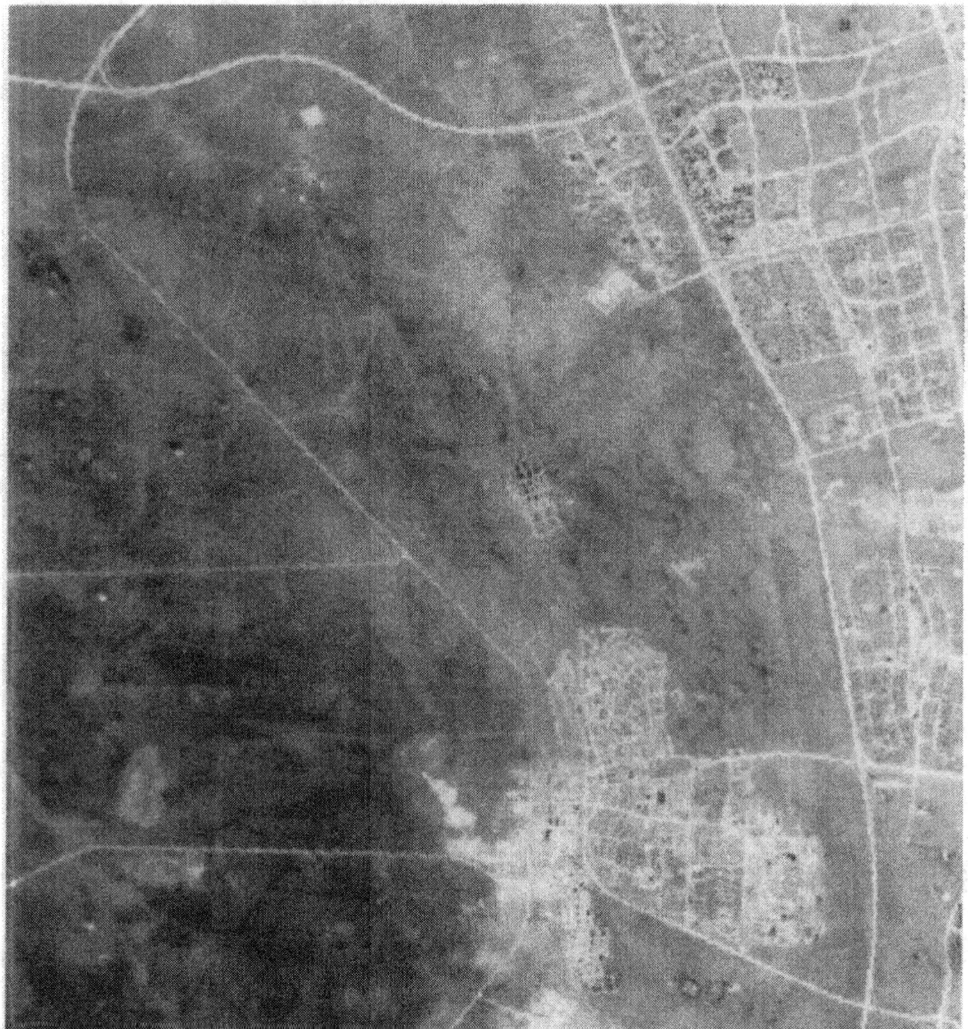

Figure 12. Change detection image produced by square root enhancement of Landsat Thematic Mapper image using band 4 (infrared; 0.79–0.9 microns) where dark pixels indicate change in the Ahmadi Oil Field of southern Kuwait.

tion, a second threshold of possible change was selected at 145. Thus, the first analysis resulted in three divisions of brightness values: 0–117 (change), 118–144 (possible change), and 145–255 (no change).

In order to determine a more appropriate threshold for the possible change segment, an additional 20 sample pixels were selected from within the town of Al-Ahmadi based on the assumption that the town itself experienced little change. Pixels were selected from areas between the roads, and the three brightness values were calculated for each of these new

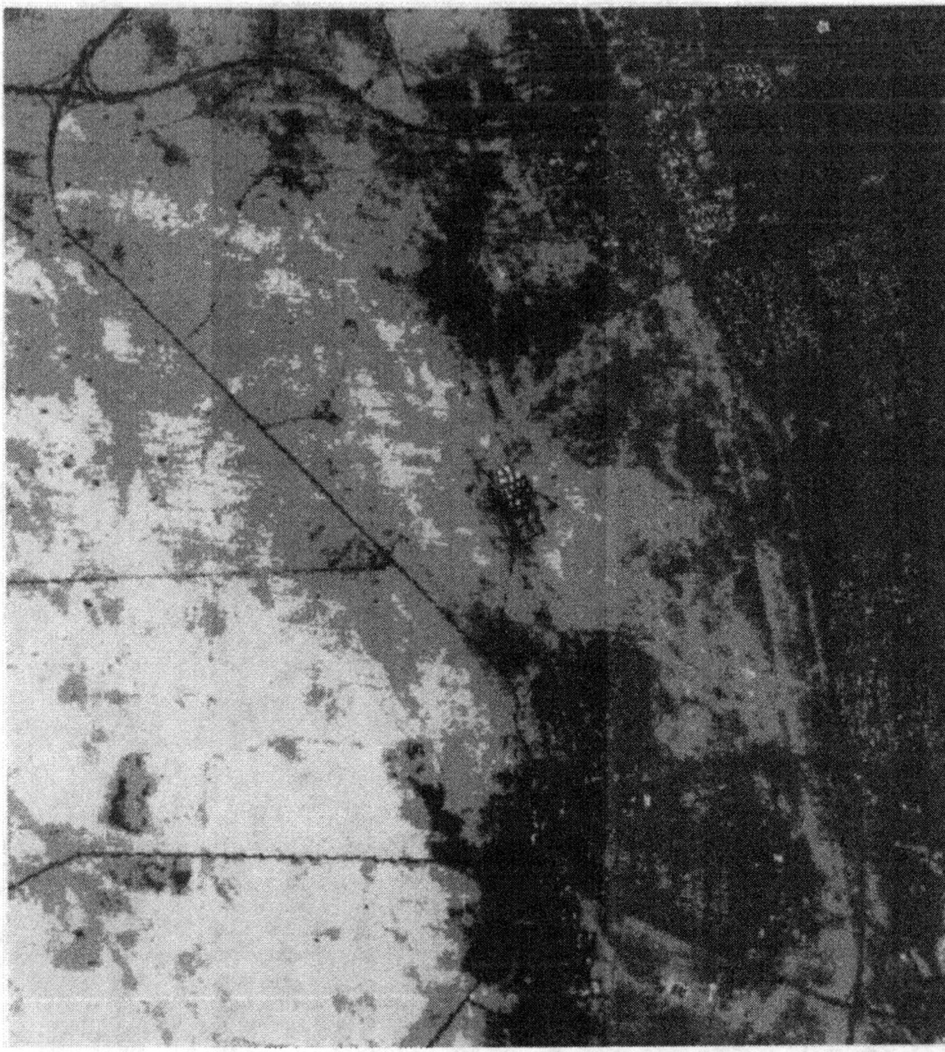

Figure 13. Bitmap image of the same area shown in Figure 12, where lighter pixels indicate change. The four grey values indicate oil-related change, any change, possible, and no change in surface characteristics.

pixels. A new threshold was set at 122, based on the brightness values of the change image, resulting in establishing the possible change range at 118–122.

The change segment needed further refinement to identify oil lakes as a distinct segment. Additional pixels were selected from those areas suspected to be oil lakes as seen on the 1992 and change image. Analysis of the brightness change of the suspected oil lakes resulted in new segments with the ranges of 0–90; and 91–117. Based on the above data, the following four categories were established for this site (Figure 13): 0–90 (oil lakes),

91–117 (deposit of oil rain), 118–122 (possible other changes), and 123–255 (no changes).

These results prove beyond doubt the utility of change detection techniques in qualitatively and quantitatively establishing the extent, and possibly the nature, of environmental change in the oil fields of Kuwait due to the Gulf War utilizing the Landsat Thematic Mapper data.

IV. GIS CORRELATIONS

For the assessment of environmental damage, Geographical Information System (GIS) methodologies are utilized in the correlation of remotely sensed information with other sources of data. GIS technology is used to provide a computer-based method of storing, retrieving, analyzing, and displaying spatially organized data. An operational GIS consists of computer hardware and software that allow use of data layers from a variety of sources. Such data include digitized aerial photographs, topographic maps, land use and road maps, satellite images, and field observations.

GIS software is used to provide a capability to: (1) store data in the most easily retrievable form; (2) update the data in a timely manner; (3) overlay various data sets for easy correlation and analysis; and (4) derive new information from existing data, which may include slight changes in the condition of the environment. Such an arrangement of data in a GIS assisted in manipulating the information for the specific needs of assessing the damage to desert surface of Kuwait due to the Gulf War, with particular emphasis on the oil lakes (Figure 14).

A set of oil lakes maps (e.g., Figure 15), produced by the Kuwait Institute for Scientific Research (KISR), were digitized and installed into a GIS database to facilitate further data manipulation. The manipulations consist, for example, of overlapping oil lakes maps on other map layers or satellite images to show the respective spatial distribution of oil lakes, and calculate number, shape, and size of lakes in each map layer. In order to perform this type of analysis, first analog map information was converted to a digital format, and then incorporated into a GIS. This process consisted of four parts: 1) map preparation, 2) manual digitizing, 3) installation of map information into a GIS, and 4) data analysis.

Photocopies of six oil lakes maps were obtained from KISR, at an approximate scale of 1:50,000, which cover five oil fields: Raudhatain, Sabriya, Maqwa, Ahmadi, and Burgan. All oil lakes were redrafted onto a clear sheet (mylar) and referenced according to a UTM grid system. Wherever possible, an original topographic map (scale 1:50,000) was used as a base map to establish the exact position of each oil lake. This was necessary because photocopies generally show some radial distortion. The oil lakes belonging to the southern three oil fields (Magwa, Ahmadi and Burgan) were manually corrected on original 1:50,000 base maps.

A digitizing software package, ROOTS, was employed to convert the oil lakes maps into a digital format. First, control points were marked on a clear mylar sheet and their respective UTM coordinates were annotated to ensure that multiple map sheets of the same or adjacent areas utilize the same coordinate system. Then, the contour of each oil lake was digitized and labeled with the corresponding category code. Finally, the digitized map was checked and edited for any errors.

Figure 14. Oil lake in the Bahra Oil Field in northern Kuwait, which is bordered by an escarpment on the north end (top; note dead bird at lower right) and the open desert to the southeast (bottom).

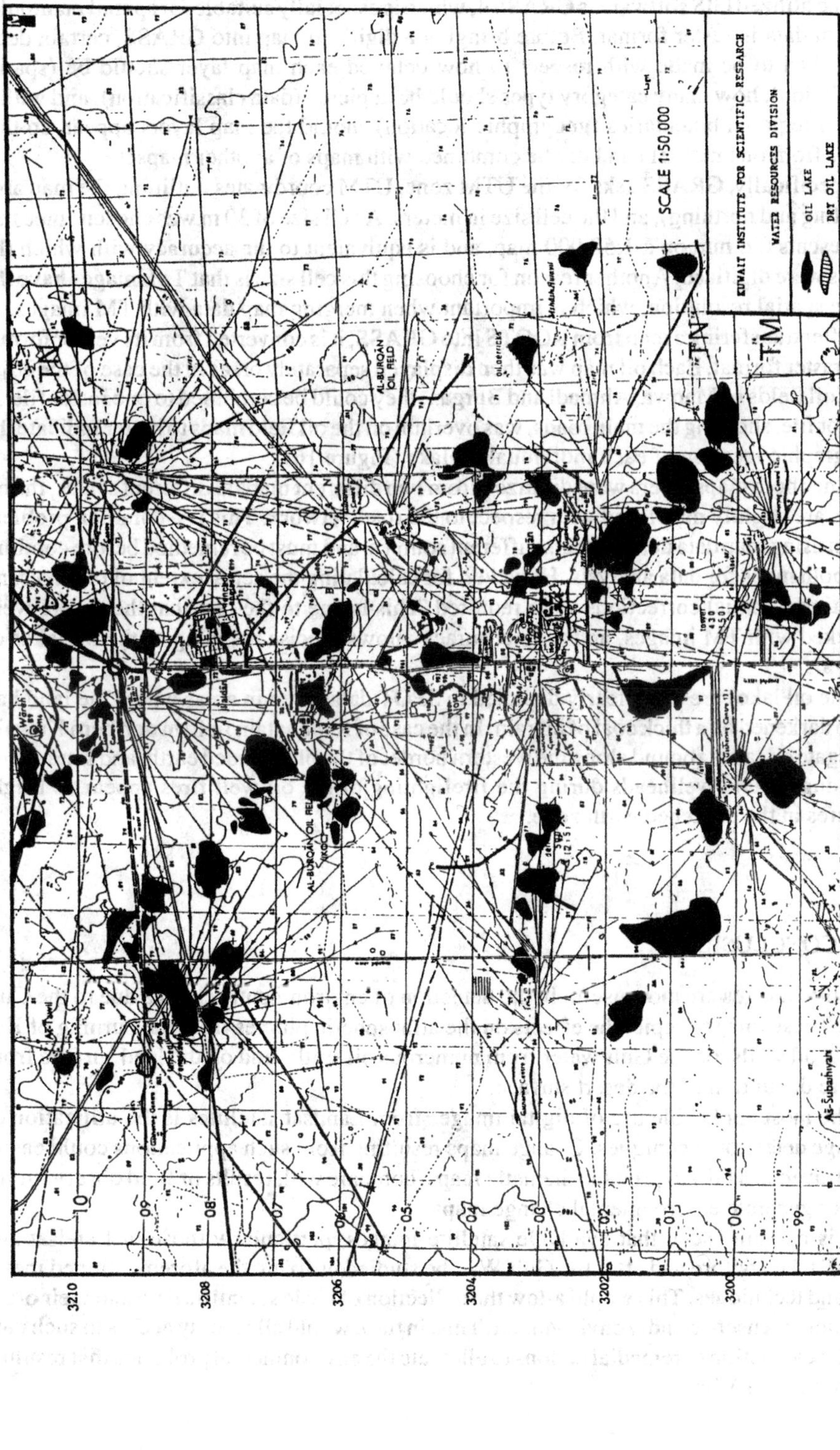

Figure 15. Map of oil lakes in the Burgan Oil Field produced by the Kuwait Institute for Scientific Research (KISR); the original was mapped at 1:10,000 scale.

The utilized GIS software is GRASS 4, which is especially suitable for spatial manipulation of data in raster format. Before bringing a digitized map into GRASS, certain decisions had to be made with respect to how detailed each map layer should be (spatial resolution), how many category types should be depicted (data classification), and within which regional boundaries (geographic location) should the map layers appear, so that maps from one mapset can later be combined with maps of another mapset.

Specifically, GRASS asks for the UTM zone, UTM coordinates outlining the map area (easting and northing), and the cell size in meters. A cell size of 30 m was chosen since this represents 0.6 mm on a 1:50,000 map, and is equivalent to the accuracy with which the maps were digitized. Another reason for choosing this cell size is that TM images have the same spatial resolution, which is important when merging map data with TM images.

After transfering a map from ROOTS into GRASS, it is converted from its vector format to a raster format. Each oil map was then displayed separately or, as in the case of the adjacent oil fields of Maqwa, Ahmadi and Burgan, they could be combined to one raster file. A vector file, showing the main roads, was overlain on the oil field maps to better indicate the geographic location of each individual oil lake (Figure 16).

Since both maps layers were digitized from two different maps at different scales, the oil lakes are slightly misplaced with respect to the road network. This is a common problem when using map data derived from different sources, and must be remedied before combining both maps for data analysis. However, for calculating the aerial extent of oil lakes, no additional spatial corrections were required. This is due to the fact that these data were checked with TM images, and their accuracy allows precise mapping of the oil lakes of Kuwait.

The oil lakes are also clearly depicted in SPOT data, even in areas where the land had been darkened by a thick coat of oil rain. In the case of Raudhatain field in northern Kuwait, elongate oil lakes abound along the western border of the blackended earth, and sand-filled pits dug around wellheads during the firefighting of the oil well fires appear as bright squares in the darkened earth zone.

V. CONCLUSION

Satellite images are most useful in the detection of environmental change due to the Gulf War. These images depict the effects on the atmospheric plumes from the burning of Kuwait's oil wells, on the Gulf water from numerous oil spills, and on the land surface from severe disruption of the desert surface.

The most applicable use of digital images from Landsat satellites is the utilization of change detection techniques. Change maps resulting from such applications could easily be utilized in a GIS to produce thematic maps, compare with results of field observations, and/or generate environmental change maps.

It is recommended that available satellite images, particularly those of Landsat and SPOT taken before and after the Gulf War, be studied utilizing the aforementioned methods and techniques. This would allow the collection of basic scientific data and their organization in a user-friendly environment. This, in turn, would allow easy access to such data for the evaluation of remedial actions to alleviate the environmental problems that resulted from the Gulf War.

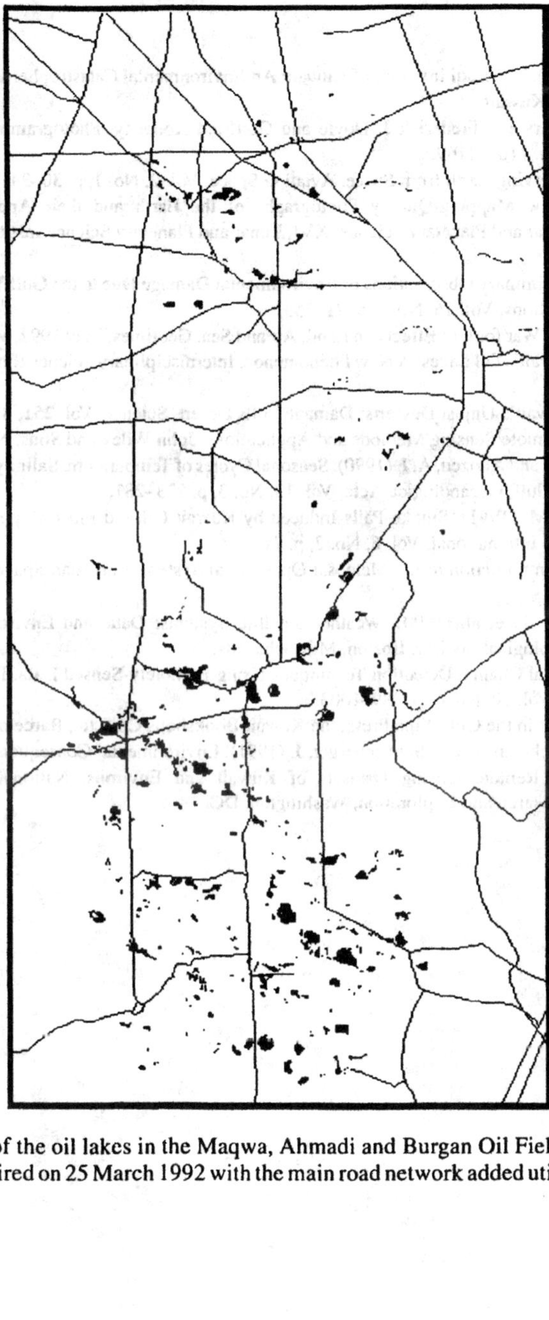

Figure 16. Map of the oil lakes in the Maqwa, Ahmadi and Burgan Oil Fields as mapped from SPOT images acquired on 25 March 1992 with the main road network added utilizing GIS methodologies.

References

Al-Hassan, J. M. (1992). The Iraqi Invasion of Kuwait: An Environmental Catastrophe. J.M. Al-Hassan, Kuwait University, Safat, Kuwait.

Doyle, F. J. 1985. Interview: Frederick J. Doyle and Gottfried Konecny. Photogrammetric Engineering and Remote Sensing, p. 1160–1169.

El-Baz, F. (1984). Observing Earth from Space. Aviation Space, Vol. 2, No. 1, p. 30–34.

El-Baz, F. (1985). New Mapping-Quality Photographs of the Earth and their Applications to Planetary Comparisons. Lunar and Planetary Science, XVI, Lunar and Planetary Science Institute, Houston, Texas, p. 207–208.

El-Baz, F. (1992a). Preliminary Observations of Environmental Damage Due to the Gulf War. Natural Resources Forum. United Nations, Vol. 16, No. 1, p. 71–75.

El-Baz, F. (1992b). The War for Oil: Effects on Land, Air and Sea. Geotimes, May 1992, Vol. 37, No. 5, p. 12–15.

El-Baz. F. (1992c). Kuwait's Oil Lakes: A New Phenomenon. Interdisciplinary Science Reviews, Vol. 17, No. 2, p. 109–110.

Holden, C. (1991). Kuwait's Unjust Desserts: Damage to Its Desert. Science, Vol. 251, 8 March 1991, p. 1175.

Hord, R. M. (1986). Remote Sensing Methods and Applications. John Wiley and Sons, New York.

John, V. C., Coles, S. L., and Abozed, A. I. (1990). Seasonal Cycles of Temperature, Salinity and Water Mass of the Western Arabian Gulf, Oceanologica Acta, Vol. 13, No. 3, p. 273–281.

Lulla, K. and Helfert, M. (1991). Smoke Palls Induced by Kuwait Oilfield Fires Mapped from Space Shuttle Imagery. Geocarto International, Vol. 6, No. 2, p. 71.

Mason, B. D. (1987). Introduction to the Meteosat Operational System. European Space Agency BR-32-ISSN 250–1589.

Rao, P. K., Holmes, S .J., et al. (1990). Weather Satellite Systems: Data, and Environmental Applications. American Meteorological Society, Boston, Massachusetts.

Singh, A. (1989). Digital Change Detection Techniques Using Remotely-Sensed Data. International Journal of Remote Sensing, Vol. 10, No. 6, p. 989–1003.

Sipa 1992. Kuwait: War in the Gulf. Sipa Press, the Kuwait Bookshops Co. Ltd., Barcelona, Spain.

Williams, R. S., Jr., Heckman, J., and Schneeberger, J. (1991). Environmental Consequences of the Persian Gulf War 1990–1991: Remote-Sensing Datasets of Kuwait and Environs. National Geographic Society, Committee on Research and Exploration, Washington, DC.

THE GULF MARINE ENVIRONMENT: VARIATIONS, PECULIARITIES AND SURVIVAL

Professor JASSIM M. AL-HASSAN

Biochemistry Department, Faculty of Science,
Kuwait University, P.O. Box 5969, Safat 13060, Kuwait

The Gulf marine environment is an aggregate of individual subenvironments. Each subenvironment is an entity on its own but interacts with and contributes to the life of other subenvironments and hence to the total Gulf marine environment. The Gulf, although small in its area, it is rather unique in its constituents and surroundings. Being a shallow body of water with the rate of exchange of its water with that of the Gulf of Oman rather slow, it can be considered as a closed lake located in a desert area. These factors make the Gulf environment rather stressed for survival, especially if the increased levels of pollution, over fishing and man's various activities in the Gulf are not checked. The Gulf environment constituents (subenvironments) are as follows:

A. MOUNTAINS ON THE COAST

The Gulf is surrounded by deserts in the west, the east, the south and the north, with some exception of Shatt Al-Arab delta in the north and Zagros coastal mountains of Iran and Masandum Peninsula at the southern end of the Gulf of the Strait of Hormuz. These mountainous coast lines are basically limestone rocks that drop suddenly into the Gulf waters forming in many locations sheer cliffs reaching a depth of over 80 ft. Under water, these cliffs act as an environment of their own. They provide anchorage to many animal and plant organisms, as well as shelter and breeding grounds to many inhabitants of the Gulf waters.

B. THE SANDY BEACHES OF THE GULF

These are commonly found in the four geographical locations of the Gulf. These beaches can be interlaced with sandstone or limestone rocks thus adding variations to otherwise a continuous line of sand. This subenvironment provides habitat for many crabs, and attract a variety of sea birds and sea turtles.

C. THE MUD FLATS OF THE NORTH

The mud flats of the Gulf span over a large area of the northern waters of the Gulf from the Kuwaiti coast in the west to the Irani coast in the east. The waters of this subenvironment are shallow and turbid.

The mud flats of Kuwait Bay, Bubiyan Island the shallows that surround Failaka Island and Abu Jazza flats are all interrelated to the flow of water from the Shatt Al-Arab into the Gulf. This component of the Gulf marine environment has a mud covered bottom with some rocks or coral outcrops and is a rich breeding and nursing subenvironment that provides some of the best commercial fish and shrimp in the Gulf. So many fish species frequent this area during their breeding migrations, while others use it as their permanent habitat. It is the subenvironment which provides the delicious fish, Zubaidi (Pampus argeneteus, Euphrasen), and one major shrimp species (Metapenaeus affinis). The waters of the mud flats are shallow and for most days of the year turbid, being the closest to the silty water of the river Shatt Al-Arab. The mud flats cover large areas of Kuwait Bay and make up the entire northern intertidal zone of the Bay. Most of Bubiyan Island is made up of wet, low lying mud flats, that are encroached upon by sea water during high spring tides and southerly gales.

D. THE INTERTIDAL AND SUBTIDAL ZONES

These subenvironments involve the intertidal mud flats of Kuwait Bay, the north coast and the intertidal and subtidal zones of the entire Gulf except the mountainous coast of the Gulf. These two zones are closely related to each other and both play a very important role in the life and bioproductivity of the Gulf. The intertidal subenvironment involves the highly productive mud flats of the north and the mangroves of the southern part of the Gulf on the eastern, the western and the southern coastlines. Even the sandy intertidal regions are highly active in their contribution towards the biomass of the waters of the Gulf. The mud-covered seabed of the subtidal zone is rich and provides feeding and breeding grounds for many fish and shrimp species, while the sandy subtidal zone is rich with fish, shrimp and sea shells, worms, crabs and sea grass beds.

E. THE CORAL ISLANDS, CORAL REEFS AND DEEP WATER SOFT CORAL BEDS

1. The Coral Reefs, Coral Islands and The Fringe Coral Reefs

Coral islands are found in the territorial waters of Kuwait and Saudi Arabia. These islands are sand covered and are surrounded by coastal fringe coral reefs. Each of these islands has a large coral reef extending in a north-westerly direction, until each reef suddenly drops into deep water as is found in the western and southern waters of the Gulf.

Although the reef surface is covered by shallow water (20–45 ft) its base is generally located in a deeper water (45–100 ft). The base is usually a hard rock which slopes into the soft sea bottom. Such reefs are found on Taylor Rock, Mudayrah, Urayfjan, the Saudi, Bahraini, Emirates and the Omani sides of the Gulf. These coral reefs represent the foundations for future coral islands. The coral reefs are made of colonies of rock building corals that are common to this part of the Gulf. The major coral types include *Acropora* (staghorn coral), *Platygyra* (brain corals), *Porites Goniastrea* (honeycomb corals) and *Turbinarias* as the major coral types. The coral reefs are an environment on their own in the middle of the vast marine environment. They are areas of mating, breeding, nursing and feeding grounds. They are habitats that act as refuge for a large quantity of sea life. The reefs support their own indigenous inhabitants as well as receive migrant and wandering fish. They are an example of communities of various animals and plants living in one habitat, exchanging benefits through some symbiotic relationship, or services rendered by one species to the other. When danger approaches, all coral community animals have an amazing capability of escaping into the clefts and crevices of the reef. The species that might protect others during a threat will find it necessary to devour other minor species for their own survival. Such a community and subenvironment is a delicate one and exists in very threatening surroundings. Water temperature is a major threat to the living reef coral community. Corals in the Gulf waters are exposed to extreme temperature variations. The summer water temperature may reach 90° F (32° C) at a depth of 80 ft., while winter water temperature could drop to 77° F (11° C). It was noted that the lowest ebb tides of the winter in the north of the Gulf could leave the upper most tips of the coastal coral reefs exposed to the freezing winds. A phenomenon like this can be seen on coastal reefs. This explains the fact that the reefs which are covered by shallow water, have dead upper coral regions. Such an environment is already at its survival limit, thus any additional environmental calamities could only further endanger the precarious situation of the coral reefs.

2. *Marine Oil Field Platforms Act As Artificial Reefs*

The waters of the Gulf contain hundreds of steel structures related to the oil industries. These include oil well platforms, sea island loading terminals and gas separation plants. These structures are found all over the Gulf, north, south, east and west. Such structures provide an excellent artificial anchorage for a vast number of animals and plants. Each structure represents a habitat for a number of seashells, sponges, crabs, fish, worms and corals. All these creatures live in harmony and exchange benefits with each other. The variety of life that constitutes the habitat of these artificial reefs varies according to the location of the structure and depth of water in which it is located.

With passing years, the biological growth on some of the structures can exceed 4 tons per structure. Thus these steel frames act as centers of attraction for migrant and roaming fish. They can also act as mating centers for these creatures. The author observed many types of fish mating in the vicinity of the platform during their mating season. Some types of fish enter spaces within the steel frame, while others swim outside the platforms. Many fish observe a zone of habitat guided by the depth of the sea, while others are wanderers where they can be found at all depths.

It is ironic that such structures which provide anchorage and refuge to such a variety of living creatures are the producers of crude oil, a material implicated in the destruction of marine life. Such wealth producing systems can be detrimental to the marine environment if badly maintained or sabotaged. The Nowruz oil field disaster is an excellent reminder of how such structures can contribute to the devastation of the Gulf.

3. *Deep Water Soft Coral Beds*

The major part of the seabed of the Gulf is either soft mud or sandy shale. This explains why the Gulf waters are excellent shrimping grounds. However, there are many locations where the sea-bed is hard rock or sandy shale with some rock out-crops. These hard bottom sea-bed areas provide an amazing environment that supports the growth of soft (non rock forming) corals, such as black corals, gorgorian, and whip corals. They also provide an excellent anchorage for pearl oysters. Usually these hard bottom areas are also rich shrimping grounds, that provide rewarding amounts of jumbo shrimp of the species (*Penaeus semisulcatus*) along with many species of deep sea fish.

Before the Iraqi invasion trawlers, through their intensive fishing practices, had almost destroyed this rich and very bio-productive environment. A corrective measure to protect the shrimp population during its breeding season by prohibiting shrimp trawling had led to improvement in the overall marine environmental situation, indicating the extensive damage trawlers inflict on this delicate and well balanced environment.

4. *Rocky Islands and Their Fringe Coral Reefs*

Such islands are found off the coasts of Qatar, the Emirates, Oman and Iran. The islands are composed of limestone and are surrounded by fringe coral reefs. Their reefs are rich with marine life and contribute actively to the life of the Gulf.

THE PRESENT ENVIRONMENTAL STATE OF THE GULF

1. *Overfishing*

Practically every country of the Gulf possesses industrial fishing fleet. Each trawler of the fleet is capable of working out at sea for a few days before it has to return to harbour. During their trawling activities a great deal of damage is done to the marine environment. The greatest damage is done to the coral head communities and to the soft coral beds. Huge areas of the brittle black coral areas have been destroyed by the effect of the trawl nets and their heavy chains. As the activities of the trawlers are not supervised, they do not abide by codes of ethics of nature preservation or environment protection, since quick financial gains are their goal. This explains the diminishing rate of catch effort in the Gulf.

2. *Industrial Pollution*

As a result to the expanding oil and petrochemical industries the Gulf is receiving a varied amounts of organic and inorganic pollutants, from these industries which are located near the coast. Although the Gulf States are much more aware of the environmental implications of such discharges (accidental or willful) and therefore more strict about the quality of what goes into the Gulf, more could be done to alleviate the levels of pollutants in the waters of the Gulf.

3. *Oil Pollution*

Oil pollution has been a part of the Gulf scene since oil was discovered in the countries of the Gulf. A new concept to oil pollution was introduced when the Iraqi military attacked

Nowruz oil field and waged the oil tanker war. The worst was still to come, when the Iraqi command decided to use the environment as a weapon of mass destruction during the Gulf war. Such intention was planned for long before Kuwait was invaded. The result of the Iraqi action is common knowledge to many people. Oil well fires also contributed a tremendous amounts of oil and soot to the Gulf waters. Their pollutants resulted from the smoke plumes of the fires.

4. Mines on the Coast, the Intertidal Zone and the Deep Navigation Channels

During the Iraqi–Irani war both side in their struggles laid mines in the waters of the Gulf. However, during the occupation of Kuwait the Iraqi military laid extensive minefields on the coast, the intertidal zone and the deep sea channel approaches to Kuwait. The coastal and intertidal mines are treacherous and will impose a serious threat to beach users and fishermen for many years to come.

WIND REGIME OF THE ARABIAN GULF

ALI HAMID ALI

Department of Geography, Boston University, Boston MA 02215

Analysis of the meteorological characteristics that influence the Arabian Gulf enhances the understanding of the major regional winds. Onset, duration, and strength of the Shamal (northwesterly) winds in winter or summer varies according to the dynamic interaction of the upper air jet streams and the distribution of the lower tropospheric pressure zones. Kous (southeasterly) winds that occur prior to the onset of Shamal winds have been characterized as being moist and warm and short lived. Severe weather such as severe dust and sand storms are associated with these winds much of the time.

Dust storms initially appear over the northwest corner of the Gulf and spread southward. It has been long noted that dust storms are associated with the development of low pressure systems over the southwestern desert of Iraq. Moreover the general synoptic conditions leading to the Shamal in summer and winter are relatively the same. However, their differences can be seen in weather conditions associated with each seasonal Shamal.

Key factors for predicting either the duration or intensity of these winds have been established by analyzing synoptic weather charts and climatological data, together with an analysis of various winds that prevail over the Arabian Gulf.

I. INTRODUCTION

The general wind meteorology of the Arabian Gulf has not been investigated in detail. While of severe consequences to the human occupants of the region, little is actually known about the meteorological characteristics of the regional winds and no efficient system for predicting either the duration or intensity of these winds has been devised. Therefore, the objective of this paper is to describe various winds that prevail over the Arabian Gulf during the year. These include three types of winds: (1) winter and summer Shamal (an Arabic word that means "north"); (2) Kous (a local term used to describe a wind from the southeast); and (3) winds that prevaile in coastal areas due to the sea breezes. The latter are not discussed in details due to their localization, while the Shamal and Kous winds are of larger scale and their effects are felt over the entire Arabian Gulf.

The analysis of Shamal and Kous winds is mainly based on synoptic weather charts and climatological data. An analysis of synoptic weather charts of certain Shamal and Kous events introduces the region's meteorology. An analysis of 13–22 years of Climatological data for some weather stations in the Arabian Gulf allows us to see how an individual station is influenced by these events.

General Climate and Meteorology

The Arabian Peninsula and Gulf region (Figure 1) are characterized by primarily dry climate that is associated with desert conditions. The general climate can be described as being a hot desert with a Koppen classification of BWh (Takahashi and Arakawa, 1981). The

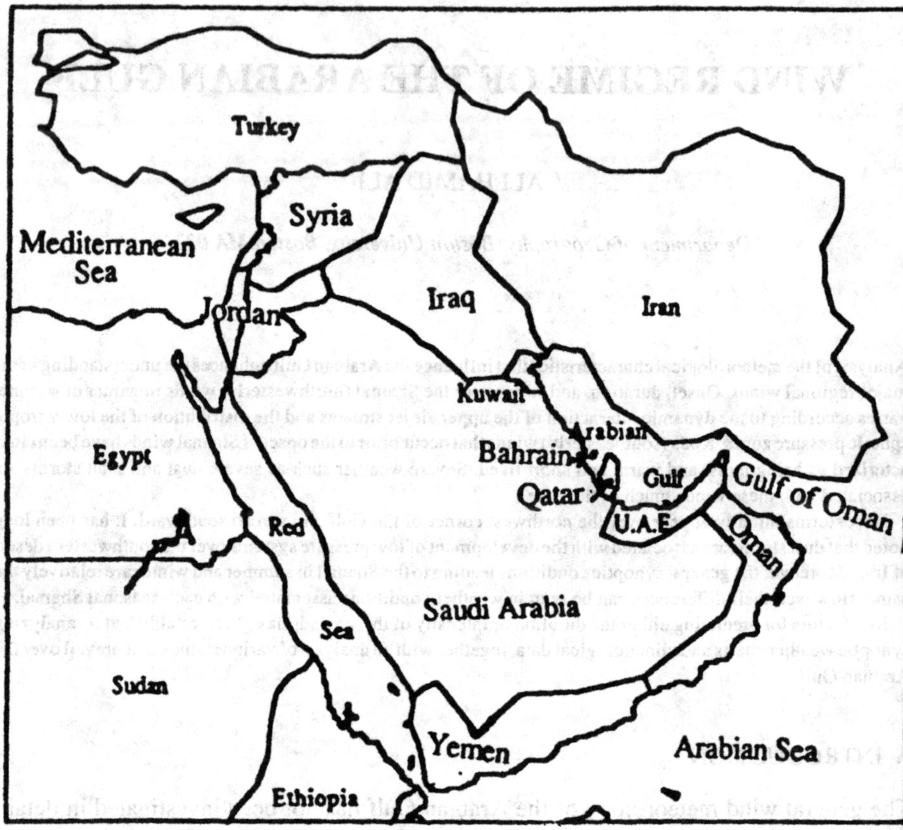

Figure 1. Map of the Arabian Peninsula and surrounding area showing the boundaries of countries in the region.

Arabian Gulf region is very hot and dry in summer and relatively cool in winter with small amounts of rainfall in winter and spring (Qatar, 1990; MEPA, 1989). The winter and spring seasons in the Arabian Gulf are very short, usually consisting of a two-month winter typically occurring in December and January, and often, only a single spring month typically occurring in March or April.

Seasonality in the Arabian Gulf is closely correlated with air temperature. Winter is characterized by mean daily air temperature below 20°C, while summer, is characterized by mean daily air temperature above 30°C. Temperatures in excess of 49°C have frequently been recorded at some Gulf stations, especially in the northern portion of the Arabian Gulf (Qatar, 1990; MEPA, 1989; Safar, 1985; AlKulaib, 1990). The temperature distribution of the Arabian Gulf throughout the year shows a range of values with the general pattern closely following orographic features and the local circulation. In winter, the temperature in the northern part of the Gulf is lower than the southern parts of the Arabian Gulf. The annual range of temperature in the Arabian Gulf region is very large, approximately 24 degree Celsius.

It has been found that in Kuwait during extreme winter nights, air temperature can drop to less than 0°C, while temperatures below –5°C have been recorded several times since 1960 (December 13, 1963 and January 1964; Safar, 1985). Riyadh, which is in a more central desert location, also experiences the same low temperature phenomena. In contrast, at Arabian Gulf coastal stations, air temperatures rarely drop to 0°C, but are commonly less than 10°C at night (MEPA, 1989).

In contrast to the cooler winter temperatures, the summer season is defined as the interval when mean daily air temperature is consistently greater than 30°C. In general during the summer, air temperatures are extremely high. The Arabian Peninsula and Gulf region are considered to be one of the hottest areas in the world (Takahasi and Arakawa, 1981).

The amount of precipitation in the region varies greatly, but a general trend of decreasing precipitation exists from north to south. The variability in rainfall is primarily due to the occurrence of thunderstorms which in general do not have a well-defined pattern of occurrence on the Arabian Peninsula and the Gulf.

Synoptic system of Arabia

In winter, the general distribution of the high and low pressure zones follow the general circulation of the global system (Figure 2). The migration of cyclones during the winter from the Mediterranean Sea into the Arabian Peninsula produces frontal systems that are responsible for rainfall in the upper part of the Arabian Gulf. While the intensification of the Indian monsoon low generates the strong Shamals across the Arabian Gulf. The association of the upper air troughs and active phases of the subtropical jet as well as the polar front jet and the migration of the Mediterranean cyclones into Arabia produce cooler winter and enhances the amount of rainfall across the region.

During the spring, which is a transitional period and time for instability in the lower level of the atmosphere, the Arabian Gulf experiences the migration of a series of depressions or troughs from the western Arabian Peninsula. These are the result of the Inter-Tropical Convergence Zone (ITCZ), which is more active at this time of year in the region near Sudan (Figure 3). At the same time cumulus clouds are developing inland. These clouds, and their associated air masses, tend to travel eastward. Moreover, spring is characterized by large daytime temperature differences between land and sea surfaces.

In summer, a thermal monsoon trough is established across the Arabian Gulf with southwesterlys over the southern edge of the Arabian Peninsula (Figure 4). The circulation pattern alters so that middle latitude disturbances of the extratropical type do not influence the area. The effect of the monsoon is also well established in the summer, increasing surface wind velocity early in the season, and increasing relative humidity in August. Surface winds start to weaken from August as the pattern weakens and the relative humidity increases near the coast. Summer conditions generally exist until late in the year and terminates with the onset of the previously described winter pattern.

II. SHAMAL WIND CHARACTERISTICS

The term Shamal is widely used around the Gulf to describe winds that are coming from the northwest and/or the north direction regardless of season. These northwesterly winds are considered the most prominent winds that prevail on the eastern Arabian Peninsula and the Arabian Gulf. The meteorological characteristics that influence the Arabian Gulf include a

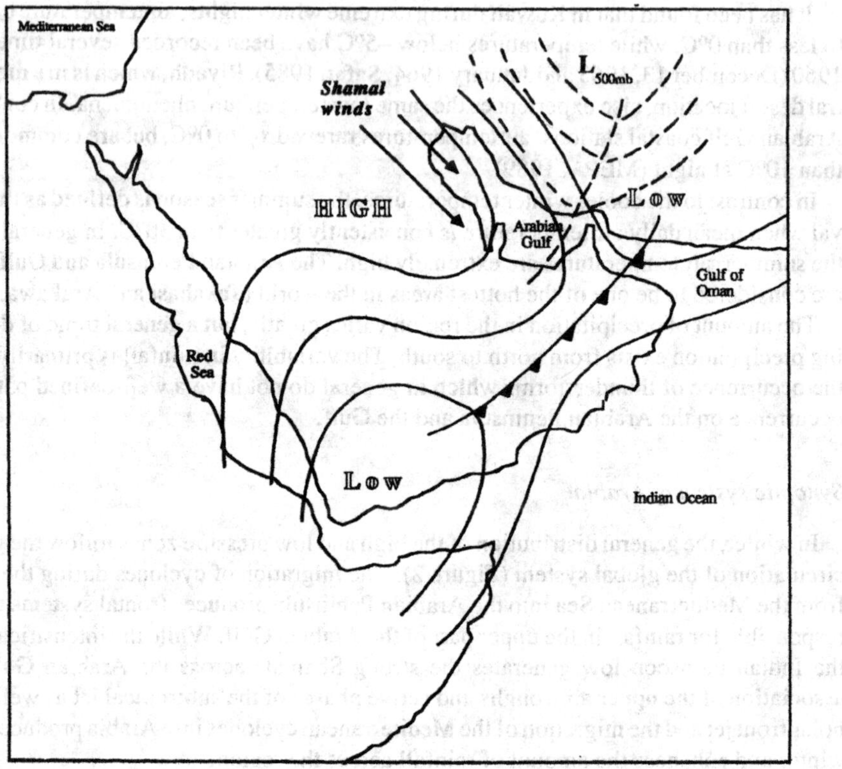

Figure 2. The direction and cause of the 3–5 day winter Shamal that persists over the Arabian Gulf. During this interval there are typically large variations in the strength of the winds. L_{500mb} indicates the low pressure system at an elevation of 19000 feet. HIGH and LOW are the pressure systems at the mean sea level.

number of regional winds that become effective at various times of the year. Other wind directions do occur such as Kous winds generally winds from the southeast direction or sea breezes on coastal lands. Occurances of a sea breeze is a good indicator of the absence of a Shamal. When a Shamal is developing, sea breeze begins to vanish. This indicates that a Shamal might prevail within 24 hours.

The usual sequence for a Shamal is the rising surface pressure over Saudi Arabia or alternatively intensification of the trough or a significant depression on surface pressure over Gulf of Oman and Pakistan. This difference in surface pressure leads to an increase in the northwesterly wind from the northern to the central part of the Arabian Gulf. However, a Shamal prevails in the southeast corner of the Gulf but becomes light and variable. The eastward movement of a Shamal usually occurs in the evening and within 24–36 hours.

After a summer Shamal, winds in mid-July and August become light and variable with the effect of sea breezes during the day becoming more pronounced as the season progresses. During this interval relative humidity frequently increases to nearly 100 percent, and at times, saturation is achieved with small cloud masses along the coastal zone. During a summer Shamal dust haze fills the atmosphere and the sky at times is partially obscured.

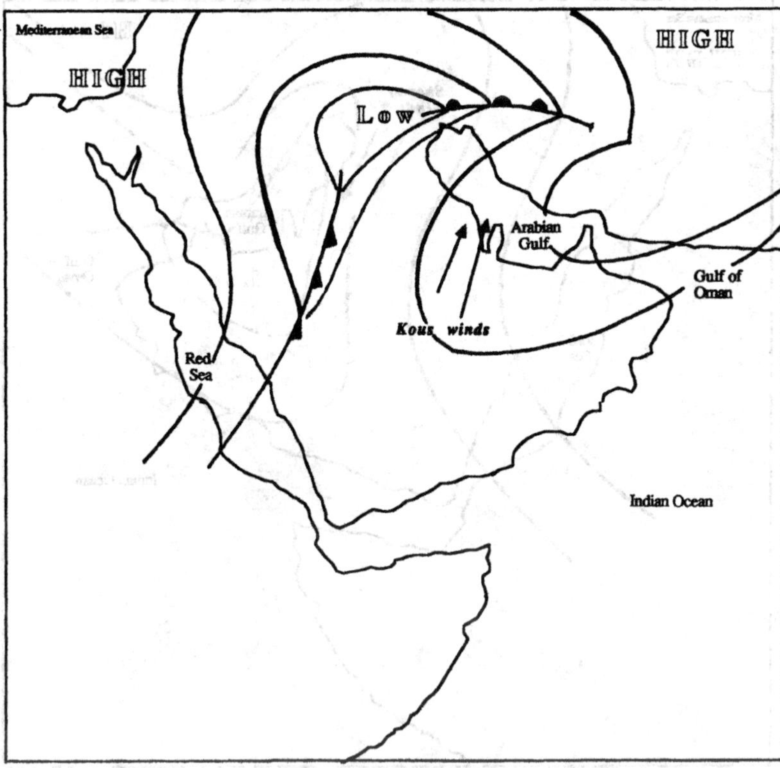

Figure 3. General spring's surface pressure distribution around the Arabian Gulf. This condition persists during March and April. Thunderstorms and Kous winds are noted during this period.

Upper Air Shamals

An important consideration of a Shamal is the strong flow in the low levels in the central Gulf. Low level jets of 60 knots or more are not uncommon at around 1500 feet above the ground whilst the wind at the ground is very light. At the 850mb level the northwesterly flow is well established all the way down the Gulf and into the Gulf of Oman (Figure 5). Another feature of the Shamal is that it blows strongly during the daylight and abate or become moderate during the night. Analysis of upper air charts shows that high pressure cells extend from the surface up to the 850 and 700 mb levels. In pre-Shamal situations, a sharp, long wave upper air trough over the eastern Mediterranean Sea remains stationary for a period of more than 24 hours due to the blocking ridges that are located over northern Africa and western and southern Turkey. Within that sharp long wave upper trough is a short wave that moves through the long wave and flattens its base. This appears to cause the long wave trough to move eastward and regroup with the Indian upper air trough. That might give a sign of formation of a surface Shamal.

The lower levels of the troposphere experience strong but shallow northwesterly wind flow during the first penetration of colder air into the Gulf region. In the upper and middle

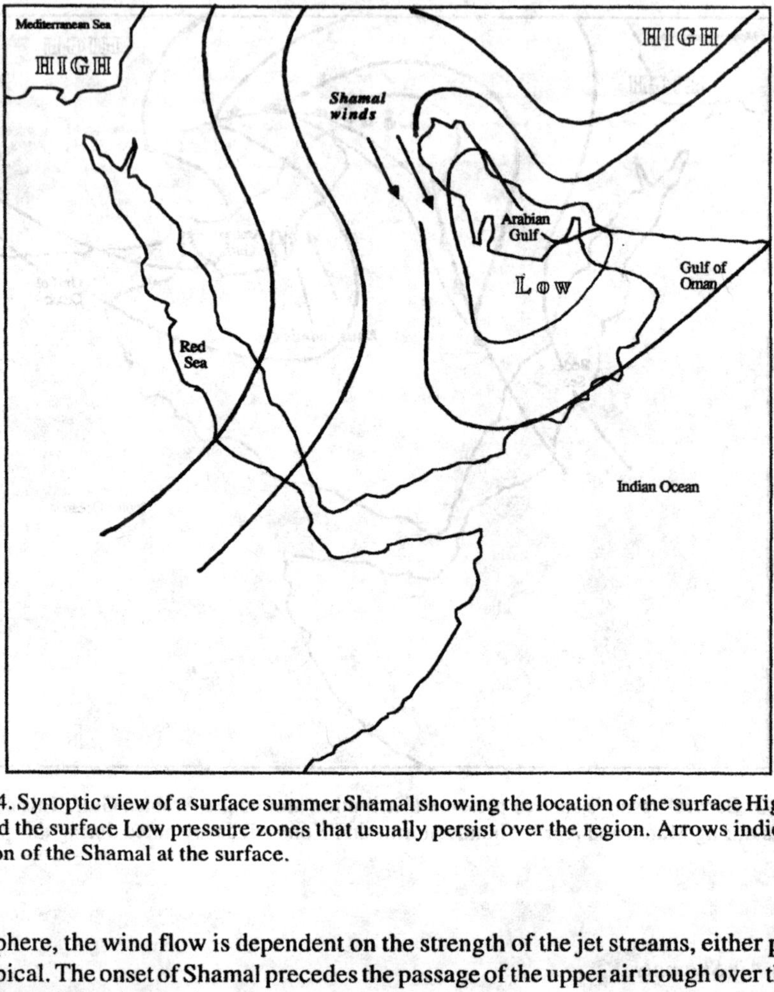

Figure 4. Synoptic view of a surface summer Shamal showing the location of the surface High pressure and the surface Low pressure zones that usually persist over the region. Arrows indicate the direction of the Shamal at the surface.

troposphere, the wind flow is dependent on the strength of the jet streams, either polar or subtropical. The onset of Shamal precedes the passage of the upper air trough over the Gulf as much as 12–24 hours.

It appears that the relationship between the surface and the upper air pressure patterns determines the duration of a Shamal (Figure 6). If the upper air trough moves quickly to the east, Shamal conditions will occur at and behind the relatively cold front as it progresses down the Gulf. However, the strength of a Shamal will decay within a short period of time after the passage of the upper trough. On the other hand, a Shamal winds will be more intense and continues for a longer period when the upper air trough becomes stationary for a time over or near the Strait of Hurmoz. However, when the upper trough moves away to the east, the Shamal weakens.

Speed and Duration

Strong and continuous northwesterly winds may occur during the winter for a period from 2 to 5 day intervals. In the summer however, the duration of the Shamal is longer with severe dust associated with the winds much of the time. During summer, for a period of more

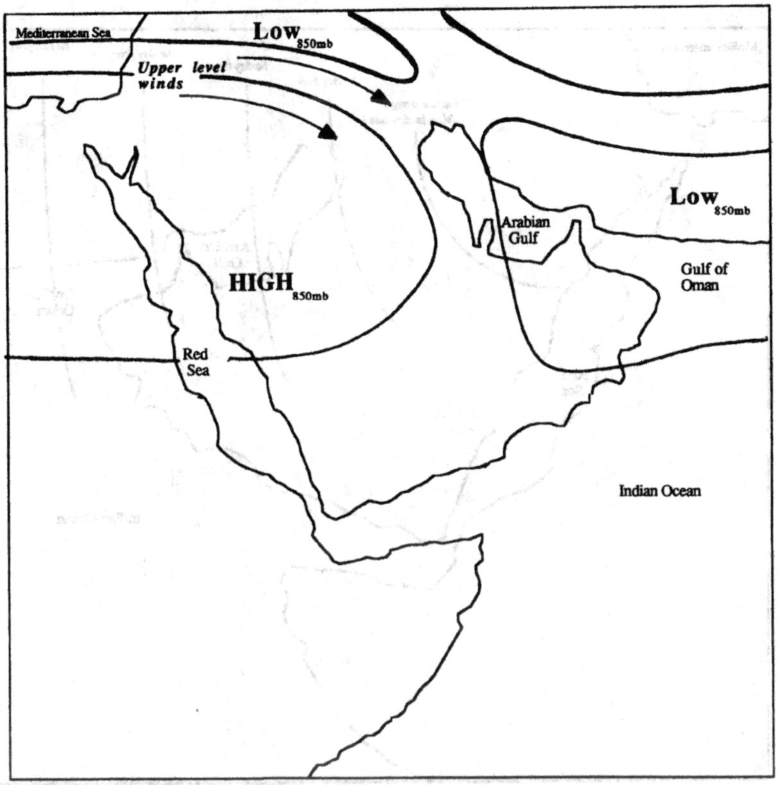

Figure 5. General upper air at 850 mb circulation (about 5000 feet above the surface) shows the general direction of a Shamal in association with the upper High pressure (**HIGH**$_{850\ mb}$) and upper Low pressure (**LOW**$_{850\ mb}$) over the region. Arrows indicate the direction of the upper air winds.

than 30 days within the months of May to July, strong winds are primarily from the northwest. Local descriptions of these winds usually carry the prefix 40-day Shamal, indicating the duration of the entire event. However, these events are rarely longer than three to five days duration, and the 40-day Shamal is really a composite of several shorter term events.

The average speed of well established Shamals ranges from 22 to 25 knots which prevail for a period of 3–5 days. In extreme cases, might extend to 13 days in a north to northwest direction with mean speed of between 12–16 knots depending on the associated upper air troughs that migrate within the easterly Subtropical Jet Streams. Shamals may become much stronger from time to time where gustiness reaches a speed greater than 60 knots in extreme Shamals.

Effects Of Topography

The Arabian Gulf is a low-lying area surrounded by mountains, including the Zagros Mountains of Iran in the east, Higaz and Assir mountains of Saudi Arabia in the west to southwest, and the Hajar mountains of Oman to the south. The region is relatively open from the north toward the Tigris-Euphrates valley. But further north, this valley is blocked

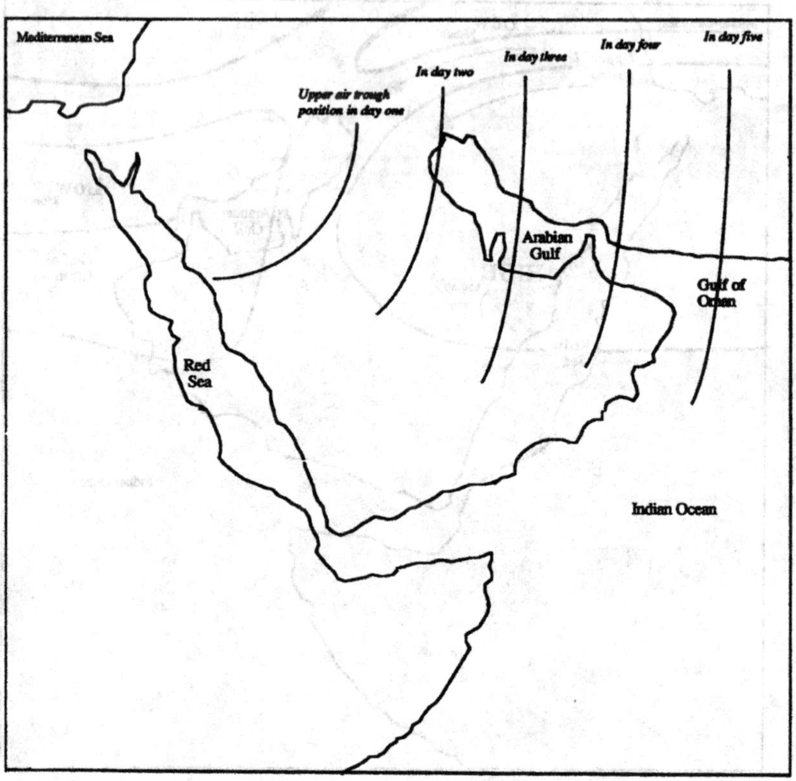

Figure 6. Progression of the upper air trough over the Arabian Gulf from day one to day five.

by the Taurus Mountain Chains in southern Turkey, the Pontic Mountains in northeast Turkey, and the Caucasus Mountains of Georgia. These mountains hieghts range from 914m to more than 3658m (Table 1). Therefore, the low-level air flow persisting over the Arabian Gulf is generally northwest to southeast as it turned around the Turkish Mountains into Lebanon and Syria then into the Arabia Gulf.

Table 1. Mountains Surrounding the Arabian Gulf.

Mountain	Location	Elevation in meters
Taurus Mts. and Pontic Mts.	Turkey	2743–3658 m
Caucasus Mts.	Georgia	2747–3658 m, some peaks are > 3658 m.
Zagros Mts.	Iran	1829–2743 m
Assir and Higaz	Saudi Arabia	914–1829 m, isolated peaks 1829–2743 m.
Hajar	Oman	914–1829 m.

The Gulf region may be divided into three parts with respect to the direction of the surface winds after the onset of a Shamal. The northwest winds which sets in after the passage of the low trough tends to be strong and gusty. However, in the northern Gulf, the general wind direction is north to west-northwest. In the middle of the Gulf, the Shamal tends to be west-northwest. While on the southeast coast of the Gulf, the prevailing Shamal direction is westerly. Near the Strait of Hurmoz, the Shamal winds blow from the southwest. It appears that the prevailing Shamal direction is strongly influenced by coastal contours (Meigs, 1966) and the effects of the neighboring mountains. Therefore, most of the Gulf stations report Shamal winds for more than 75% of the times while the remaining directions are due to the sea breezes or local effects (Figure 7).

III. WINTER SHAMAL

An onset of a typical winter Shamal occurs after the passage of a cold frontal trough from eastern Mediterranean Sea into Arabia. The passage of this front reaches areas near Kuwait. The eastward movement of this cold front is due to the intensification of a high pressure zone that enters Arabia from the west and the falling pressure ahead of this frontal system. A development of pressure gradients enhances the speed of surface winds which usually persists for a period from 2 to 5 days. The movement of the surface trough over the Arabian Gulf is southeast ward, which follows the pattern of decreasing pressure over the Gulf of Oman and the intensification of semi-permenet high pressure over northwestern Saudi Arabia.

A typical winter Shamal development is presented in Figure 8. An upper air trough acroses eastren Mediterranean Sea and acts on the surface low pressure over Syria and Jordan. The upper air trough will cause the low pressure trough over the Red Sea to move toward Oman which causes warm and moist south to southeast surface winds over the Gulf (locally called the Kous winds, and will be discussed later). The movements of these, the upper air trough and the surface low, is ususlly to the east. Most of the times a cold air mass fall behind the surface trough and cumulus type clouds precede this trough usually to the north of the Arabian Gulf. Thunderstorms and precipitation might occur over the northern parts of the Gulf. The eastward movement of the surface trough from the Red Sea is followed by the entrance of an African high into central Saudi Arabia. Surface pressure continues to rise as it accelerate the surface trough and the associated low eastward. Subsidence in the lower troposphere induces a surface high pressure area over northern Saudi Arabia. An increase in surface Shamal begins over most parts of the Gulf as the upper air trough moves eastward. Figure 8 shows that the upper trough to eastern Iran had advanced the surface low. The associated cold front had swept down the Gulf and into the Arabian Sea. The mechanism between the increasing high pressure over central Saudi Arabia and the low over the Gulf of Oman produces a strong Shamal over the Gulf. As subsidence continues within the high pressure over Saudi Arabia in the lower troposphere.

IV. SUMMER SHAMAL

The area on which a summer Shamal blow extends from southern Iraq down to Kuwait, the eastern and central provinces of Saudi Arabia, Bahrain, Qatar and up to the United Arab Emirates, only Khasab and Buraimi of Oman have been found to be affected by the Shamal. Inland, the strong winds reach a distance of 250 miles from the Gulf coasts. They are

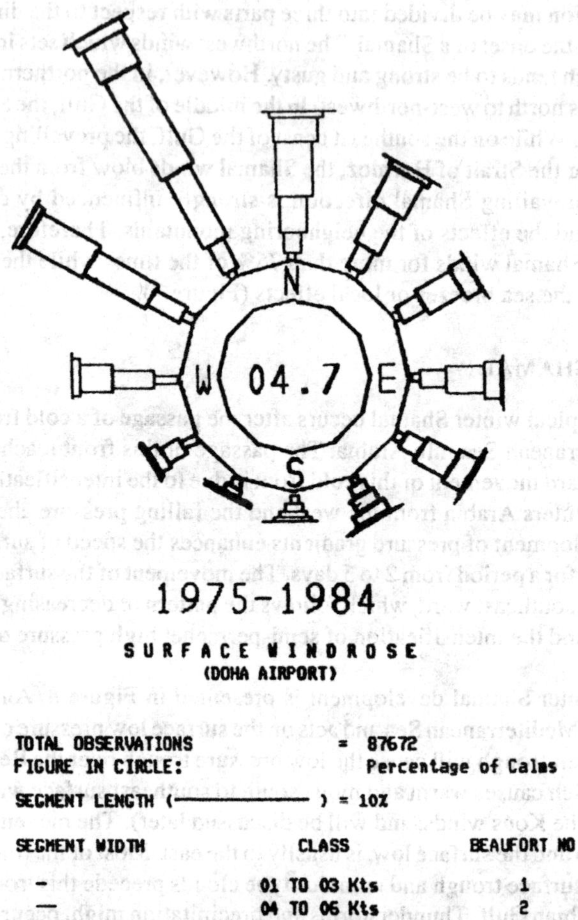

Figure 7 . Surface wind rose at Doha International Airport weather stations from 1975–1984 depicting the Shamal and Kous winds.

strongest on the west side of the Gulf and fade away gradually twards the east, so that they die out upon reaching the Iranian coast.

The expansion of the Indian monsoon low (IML) into the Arabian Peninsula forms a large area of low pressure over the Arabian Gulf. The lowest pressure zone usually over Oman, while semi-permenent high pressure zone persist during the summer over northwestern Saudi Arabia. During this time the Gulf waters act as a source of sensible heat. Therefore, a thermal low establishes over the southeastern portion of the Gulf. As a result, a thermal trough becomes as a part of the IML trough which extends from the northern Ara-

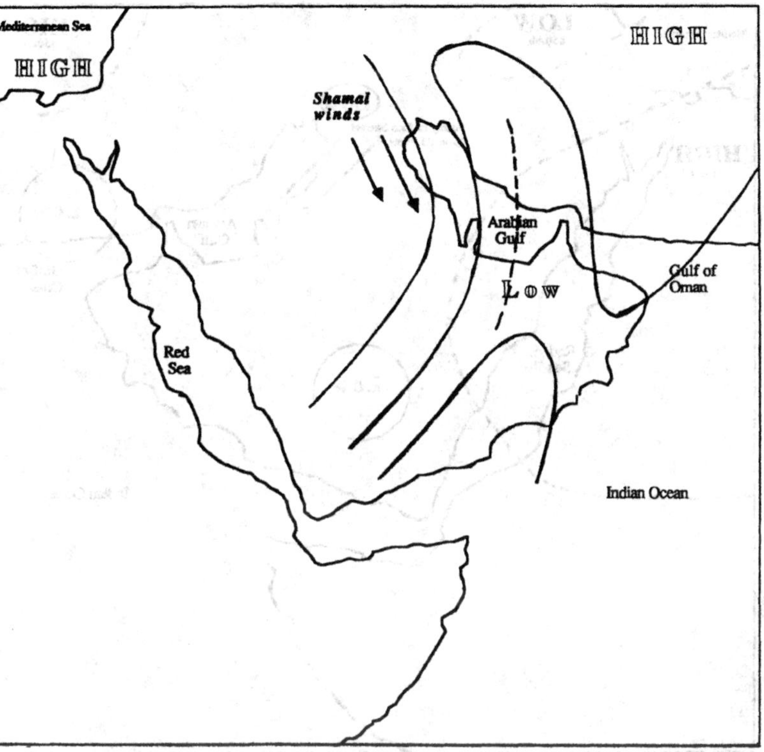

Figure 8. An upper air trough pentrates into Arabia from the Mediterranean Sea with an easterly movements. The surface Shamal winds begin to dominate from 3 to 5 days over the region with an average surface wind speed of 25–30 knots. Dashed line indicates the location of the upper air trough at 19000 ft (500mb).

bian Sea into Iraq. Strong pressure gradients develops along the Arabian Gulf with mean sea level pressure variation ranges 4–8 millibars between Gulf of Oman and central Saudi Arabia. This pressure variation enhances the Shamal winds, during the day, up to 37 knots at the surface (Figure 9).

Shamal wind velocity decreases during the night at low levels of the atmosphere but at higher altitudes winds remain very strong. A typical low-level profile from reports received during a notable Shamal outbreaks of June 1980 are plotted by Membery (1983). The wind profile shows a remarkable uniformity in wind direction. Surface wind speed was 16 knots but at an altitude of 300 meters the Shamal winds were averaging 45 knots. At higher levels (approximately 1000 meters) the Shamal was still stronger than at the surface but had weakened considerably from speeds experienced at 300 meters.

Similarities and Differences Between Summer and Winter Shamals

The general synoptic conditions leading to the Shamal in summer and winter are the same. For example, in both the development of high pressure over northwestern Saudi Arabia

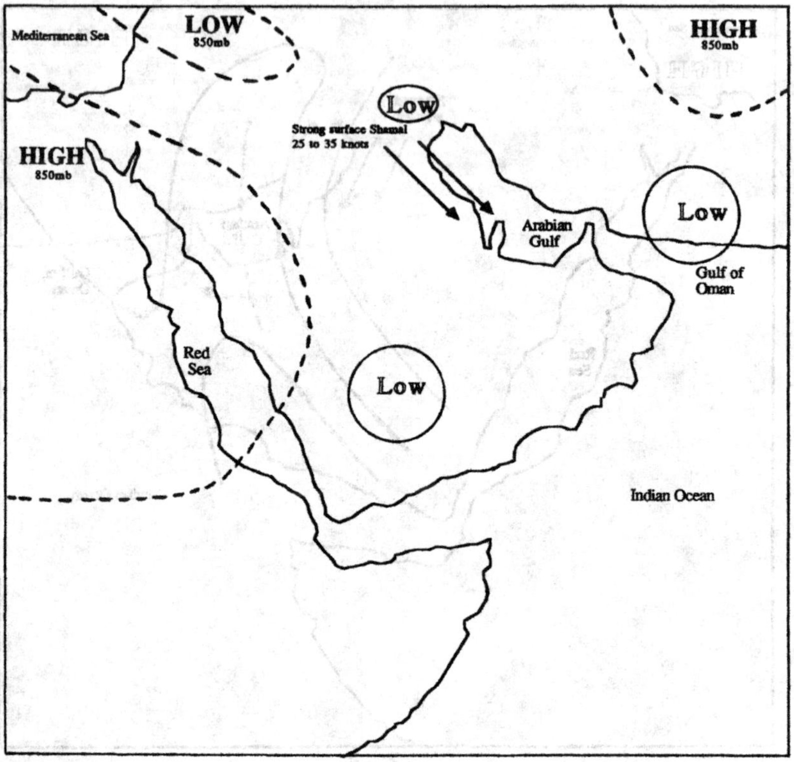

Figure 9. Location of upper air high pressure and low pressure systems at an elevation of 5000 feet above the surface (**HIGH$_{850mb}$** and **LOW$_{850\ mb}$**) over the region with respect to the surface low pressure would result an average wind speed of 25 knots.

and the intensification of the low pressure over the Gulf Of Oman. However, their differences can be seen in weather condition associated with each Shamal Onset. For example, in winter low level clouds are present ahead of an approaching low pressure trough. Rainfall tends to occur more over the northern parts of the Gulf than the southern parts, especially during the months of November and April. Decrease in air temperature over the Arabian Gulf always follow the prevalence of northwesterly winds. However, the summer Shamal is more likely to be dry and have the potential to create dust and sand storms more than the winter Shamal. High cirrus clouds appear over the coastal regions of Oman in association with the penetration of the IML.

As stated earlier, a summer Shamal may occur with the presence of the high pressure cell over central Saudi Arabia and development of the low pressure trough over the Gulf of Oman. However, the winter Shamal develops with the passage of a low pressure trough that enters Arabia from the Mediterranean Sea. The pressure over northwestern Saudi Arabia tends to increase as a high pressure cell from northern Africa moves toward the Arabian Peninsula. As the fall in pressure increases over the Gulf of Oman, a steep pressure gradient forms, resulting in an increase in the wind speed from the northwest direction. There is also an upper air jet stream (Sub-Tropical Jet Stream) that tends to create a deep upper air trough. The deepening in that trough increases the Shamal wind velocity at the surface. This situation is more common in winter Shamals.

A migration of low troughs and high ridges over the Arabian Peninsula from the northwest normally starts during the months of December, January and February. During the months of March and April, troughs usually migrate from western Saudi Arabia into the Arabian Gulf.

V. KOUS WINDS

The term Kous is locally applied to any wind from a southeasterly direction. In general, these occur prior to the onset of Shamal winds. The passage of the low pressure systems from the eastern Mediterranean sea into northern Arabia causes these Kous winds to occur. The Kous winds are characterized by being moist and near the coastal areas but dryer with higher temperatures inland. They are known for bringing dust from the central desert of Saudi Arabia into the Arabian Gulf. Visibility diminishes to less than 5 km and at times to less than one kilometer in severe dust storms. Since Kous winds are associated with the passage of a frontal system during the winter, their duration ranges from 20 to 48 hours. Their wind speed reaches 20 to 30 knot, but varies with respect to the passage of the low pressure trough in a southeast movements from northern Gulf, mainly during late September until early May.

A general synoptic pattern for Kous winds during the winter would be considered as a pre-Shamal situation with the entrance of the low pressure from eastern Mediterranean Sea and become stationary for at least 24 hours over Syria. At that time, a cold air would approach the northwestern portion of Saudi Arabia. The air temperature may fall to less than 7^{0}C as recorded in Hail on December 22/1971. The low over Sudan would act as a source of warm air while a high pressure system that is centered over central Asia brings colder air to the Arabian Gulf. The wind direction of these two later pressure systems is from south to southeast component. The southeasterly wind direction would be prevalent over the entire Arabian Gulf of a period of 24–48 hours. The main reason for this prevalence is the low pressure system that is stationary over Syria and the Asian high pressure system. This low system has the potential to move northeastward within 20 to 48 hours depending on the active phase of the upper air short wave as described earlier in winter Shamal (Figures 8). An example of this synoptic situation is presented in Figure10 (from Al-Kulaib, 1990, p.114). Note the surface contours near the Arabian Gulf and the direction of winds they produce.

Another type of Kous winds that prevail over the central portion of the Arabian Gulf during the period from early March until mid-May. This type of Kous would be a product of a local low system that become stationary over central Arabian Peninsula. The Kous winds will exist until areas north of Bahrain and is characterized by being warm and dry at the southern Arabian Gulf. Areas near Kuwait and at the regions ahead of the warm front would experience east to northeast winds then followed by northwesterly winds. An example of this synoptic situation is presented in Figure11 (from Al-Kulaib, 1990, p.115). Note the surface winds at the Arabian Gulf and central Saudi Arabia.

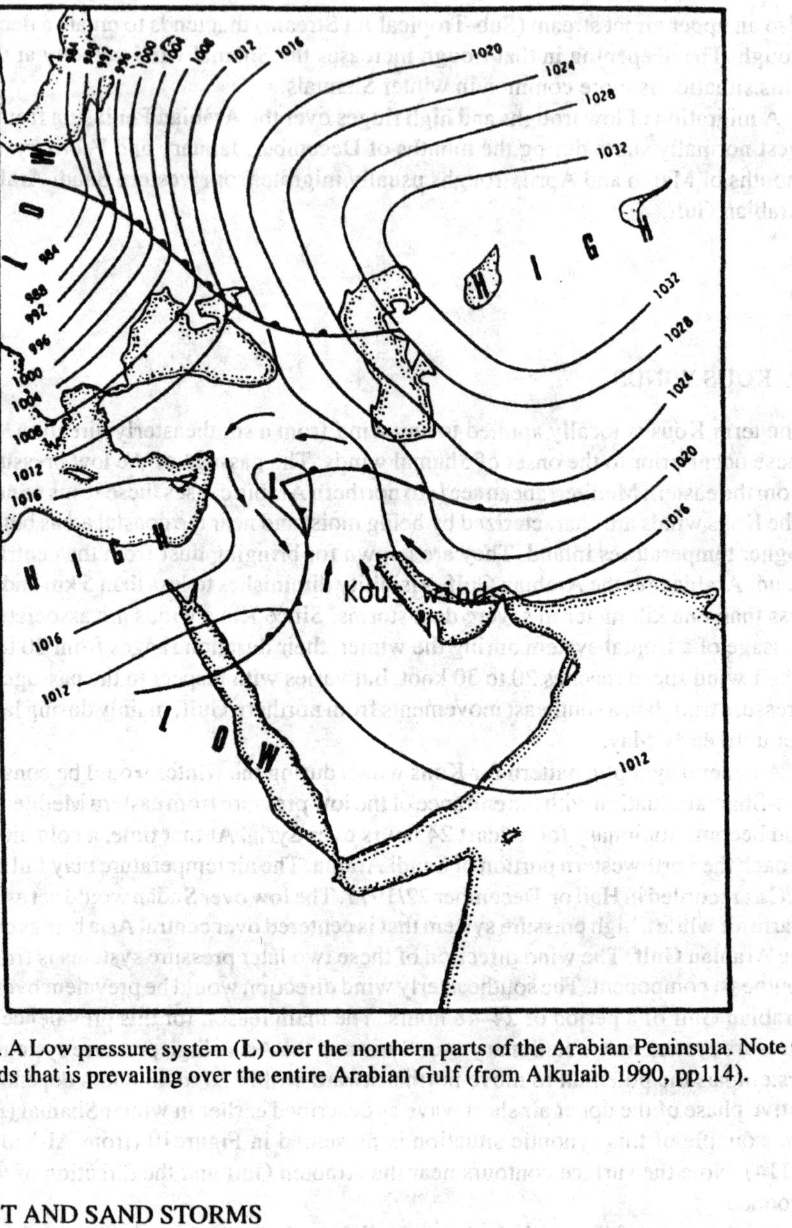

Figure 10. A Low pressure system (L) over the northern parts of the Arabian Peninsula. Note the Kous winds that is prevailing over the entire Arabian Gulf (from Alkulaib 1990, pp114).

VI. DUST AND SAND STORMS

It has long been noted that dust storms in the Gulf region will usually begin whenever a low system develops over southwestern Iraq. The thickness of the entrained dust depends on the vertical motion of the two pressure systems near that region. It normally takes about 12 to 24 hours for the dust to spread over the Gulf with the timing primarily dependent on the strength of the wind flow and the upper air jet streams.

Key factors that lead to the development of cyclogenisis which leads to the initiation of dust storms from southwestern Iraq to the northwest portion of the Gulf, and along the en-

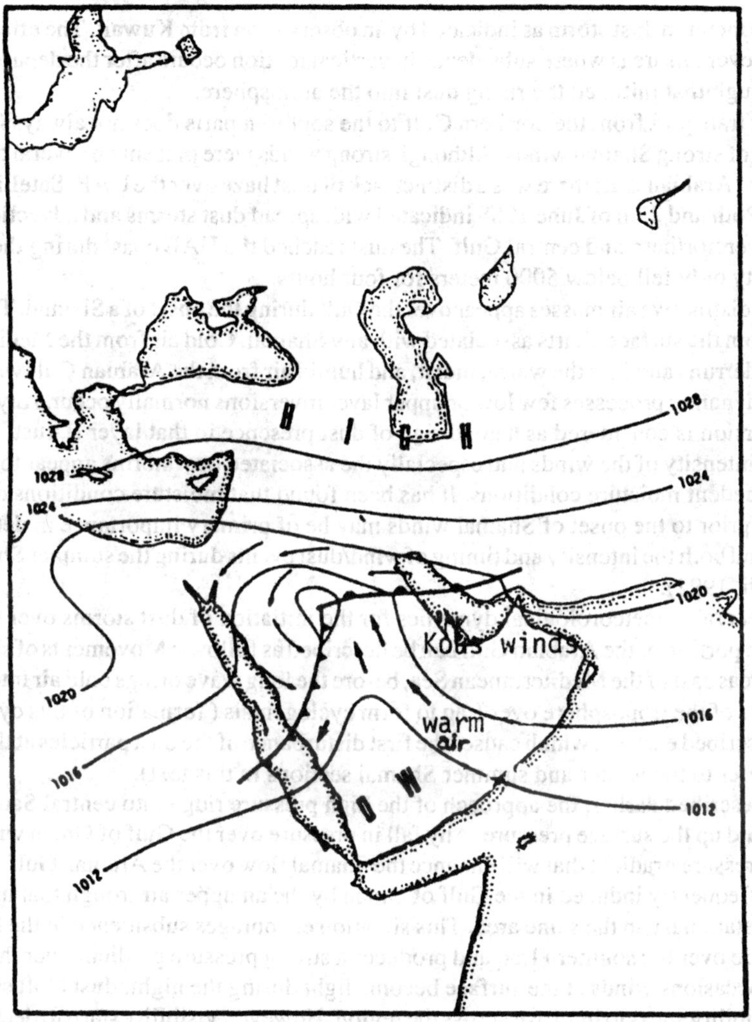

Figure 11. A local Low pressure system (L) over the central parts of the Arabian Peninsula. Note the Kous winds that extends to the central Arabian Gulf. Areas in the northern parts of the Gulf are enfluenced by easterly winds (from Alkulaib 1990, pp115).

tire Gulf include; 1) warm Gulf waters which are considered to be a source of sensible heat, 2) ascending vertical motion within the low trough, and (3) convergence of surface flow between the southeasterly winds (Kous winds) and an approaching trough that has northwesterly wind flow behind it as a result of high pressure subsidence.

An example of severe onset of a summer Shamal was noticed over the Arabian Gulf in June 14–1989 (Ali and Scuderi, 1992). That Shamal had covered the Arabian Gulf with an average surface winds of 20 knots. Visibility during that Shamal had reduced to less than 300 meters on the northern part of the Gulf. Areas in the northern portion of the Gulf experience dust storms and dust haze. In severe storms the horizontal visibility decreases to less than 50 meter in dust storm as indicated by an observation from Kuwait. The effect is especially severe in areas where subsidence in vertical motion occurs after the departure of the low trough that initiated the rising dust into the atmosphere.

Dust transport from the northern Gulf to the southren parts does not always follow the pattern of strong Shamal winds. Although strong winds were present on several days in the northern Arabian Gulf there was a distinct lack of dust haze over the UAE. Satellite images on the 26th and 27th of June 1988 indicated widespread dust storms and advection of dust haze over northern and central Gulf. The dust reached the UAE coast during the 27th but visibility only fell below 5000 meters for four hours.

Two distinctive air masses appear over the Gulf during the onset of a Shamal. This can be seen from the surface charts associated with any Shamal. Cold air from the Mediterranean Sea underruns and lifts the warm, moist, and humid air from the Arabian Gulf waters. Due to this dynamic processes few low or upper layer inversions normally occur. Any presence of inversion is considered as a good sign of dust presence in that layer (Drust, 1935).

The intensity of the winds and especially the associated dust storms appear to be linked to antecedent moisture conditions. It has been found that moisture conditions during the season prior to the onset of Shamal winds may be of primary importance in determining onset and both the intensity and timing of wind/dust events during the summer Shamal season (Ali, 1992).

In general, a meteorological dynamics for the initiation of dust storms over the northwestern portion of the Arabian Gulf can be described as follows; Movements of short wave from areas east of the Mediterranean Sea, before the long wave brings cold air into the lower layers of the troposphere over Iraq to form cyclogenisis (formation of this cyclogensis was described earlier), which causes the first disturbance of the dust particles at the surface (also refer to the winter and summer Shamal sections in this text).

As described earlier, the approach of the high pressure ridge into central Saudi Arabia will build up the surface pressure. Any fall in pressure over the Gulf of Oman will create a steep pressure gradient that will enhance the Shamal flow over the Arabian Gulf. A surface low is frequently induced in the Gulf of Oman by the an upper air trough that usually becomes stationary in the same area. This situation encourages subsidence in the lower troposphere over the southern Iraq and produces a strong pressure gradient over the Gulf. In many occasions winds at the surface become light during the night, dust aloft start to fall and visibility starts to improve across the region. However, visibility startsto decrease during strong Shamal events from midday until the evenings.

VII. CONCLUSIONS

The Shamal and Kous winds over the Arabian Gulf are the result of the dynamic interaction of the upper atmosphere and the physical climate of the surface. The general synoptic view of the region shows that an advance of a high pressure ridge into central Saudi Arabia and intensification of a depression over the Gulf of Oman are the key factors in strengthening a Shamal over the Arabian Gulf. An increase in surface temperature may enhance the velocity of meso-scale winds. However, large differences in mean sea level pressure around the Arabian Gulf may influence the macro-scale winds. As a result, strong pressure gradients over the Arabian Gulf direct and accelerate both surface winds and winds in the lower 2500m of the troposphere. At the same time, the subtropical jet stream influences the wind flow at middle and upper troposphere levels. Gustiness of Shamal winds are associated with a deepening of the upper air trough that is induced by the subtropical jet stream. Moreover, the topography of the Arabian Gulf coast is a factor in directing surface Shamals. In addition, there is evidence that the Shamal weakens at night while remaining strong aloft, and sea breezes prevail over the Arabian Gulf in an absence of a Shamal, while Kous winds are considered to be a pre-cursor for the Shamal in general.

The duration of the summer Shamal is longer than the similar winter Shamal event, even though the meteorological dynamics in and surrounding the area are the same. It appears that the relationship between the surface and the upper air pressure patterns determines the duration of the Shamal in both seasons. During the winter Shamal, the location of the upper pressure trough over the Gulf of Oman determines the duration of the onset. Short Shamals are associated with air troughs that last 12–24 hours, while 3–5 day duration Shamals persist as long as the upper air trough remains over the Gulf of Oman. Summer Shamal duration is primarily associated with the movements of a high pressure ridge from Africa into Arabia, and the subtropical jet stream at both the 850 and 700 mb levels. Increases in surface pressure are primarily related to the African high as it penetrates into the Arabian Peninsula. A decrease in surface pressure over the Gulf is primarily related to the Indian monsoon low. The interaction of these two pressure systems sets the duration of the summer Shamals.

The duration of Kous winds is primarily related to the movements of the northern and central low systems that penetrates into the Arabian Peninsula from the Mediterranean Sea and is very short lived. Kous winds primarily occur during the months of September to early May. The Arabian Gulf experiences two kinds of Kous winds, one over the entire Gulf, and the second over north of Bahrain. Both of these Kous winds bring warmer and both dry and moist air into the area. Both Kous winds bring dust storms from the open plains of central Saudi Arabia.

Dust and sand storms that prevail to the west and over the Arabian Gulf are primarily originated in southwestern Iraq. Satellite images indicate that certain points in the Iraqi desert and to the west of Euphrates River act as a source for the dust storms. Strong Shamals of 20 knots or more are required to initiate entrainment of this dust and to transport it southward.The transport of the dust from the northern to the southern part of the Gulf depends on several factors related to the meteorology and geomorphology of the regions traversed by the dust. Dust storms occur after specific major factors are satisfied, including dryness of the soil or air, strong winds, and high surface temperatures.

It appears that several factors are key to contorting the onset, duration and intensity of a Shamal. Further studies, with a larger and more regional meteorological grid should be directed towards investigating the accuracy of these findings. Furthermore, winds in the Gulf region are influenced by the general circulation of the atmosphere and therefore changes in the characteristics of Shamals and Kous winds may be indicative of larger global-scale climatic change.

References

Ali, A. H. 1992, Wind meteorology of the summer Shamal in the Arabian Gulf region. Unpublished Master thesis, Boston University, Geography Department.

Al-Kulaib, A. A. 1990, Weather and Climate of the State of Kuwait. Kuwait Department of Meteorology, Kuwait.

Ali, A. H. and Scuderi, L. A. 1992, Dust intrusion events related to the summer Shamal in the Arabian Peninsula. In preparation. Department of Geography, Boston University.

Drust, C.S. 1963, Dust in the atmosphere. Quarterly J. of the Royal Met. Soc.(61): 81–89.

Meigs, P. 1966, Geography of coastal deserts. Paris, UNESCO Research.
Membery, D. A. 1983, Low level wind profiles during the Gulf Shamal. Weather, 38 (1): 18–24.
Qatar Department Of Meteorology. 1990, Long period means & extremes of climatological elements for Doha International Airport, (Period:1960–1990). Climate Section Publications. Doha, Qatar.
Safar, M. 1985, Dust and sand storms in Kuwait. Kuwait Department Of Meteorology. Kuwait.
Saudi Arabia Meteorology & Environment Protection Administration (MEPA). 1989, Yearly surface Climatological Data. Means and Extremes Of Selected Weather Elements.
Takahasi, K. & Arakawa, H. 1981, Climate of southern and western Asia. In World Survey of Climatology.

MODELING OF AIR CURRENTS IN THE GULF REGION*

THOMAS J. SULLIVAN, JAMES S. ELLIS, CONNEE S. FOSTER, KEVIN T. FOSTER, RONALD L. BASKETT, JOHN S. NASSTROM and WALTER W. SCHALK

The Atmospheric Release Advisory Capability
Lawrence Livermore National Laboratory
University of California
Livermore, California 94551–0808

The Atmospheric Release Advisory Capability modeled the wind flow in the Gulf Region in order to make projections of the Kuwait oil fires pollution dispersion. Extensive meteorological models incorporating explicit terrain influences to the flow fields were routinely employed through a six month international assessment support effort organized by the World Meteorological Organization and U.S. scientific research agencies. Results show generally close agreement with visible imagery of the smoke plumes as detected by meteorological satellites. However, there are some examples of significant disagreement or failure of the meteorological models. These failures are most likely directly linked to missing or unavailable weather observations.

INTRODUCTION

The Atmospheric Release Advisory Capability (ARAC) was originally conceived and developed at Lawrence Livermore National Laboratory as a nuclear accident emergency response service.[1] It's purpose is to provide near real-time dose assessment calculations for accident response decision makers. Based on operationally-robust three-dimensional atmospheric transport and dispersion models, extensive geophysical and dose factor databases, real-time meteorological data acquisition and highly experienced staff, ARAC responds to radiological accident events in the United States within 30–90 minutes. Through extensive research and experience, ARAC has demonstrated the capability to quickly adapt its modeling system to any accident location.[2]

Beginning with the Chernobyl accident, ARAC has, on several occasions, been requested to calculate the transport and dispersion of hazardous substances for accidents outside the United States.[3] One of the latest such requests was for support of the U.S. research aircraft flight programs measuring the pollutants from the Kuwait oil field fires (May 12 to June 15 and July 26 to August 20, 1991). This was followed by a United Nations World Meteorological Organization (WMO) request for assistance to the Gulf region's environmental and meteorological services. Beginning on May 12, 1991 ARAC initiated calculations of the wind flow and pollutant dispersion conditions of the Gulf region which continued

*Work performed under the auspices of the U.S. Dept. of Energy by the Lawrence Livermore National Laboratory under contract No. W-7405-ENG-48.

through October 31, 1991 just 6 days before the last well fire was extinguished. These calculations were produced for a 3200 kilometre square region using meteorological data sources from the U.S. Air Force Global Weather Central (AFGWC) in combination with the ARAC wind field and pollutant dispersion models which included effects of terrain. AFGWC provided both analysis and prognostic gridded datafields from a Relocatable Window Model (RWM) which produced wind component data on a 40 x 40 grid at 50 nautical mile spacing at 1,000 ft, 2,000 ft, 5,000 ft and every 5,000 ft up to 30,000 ft elevation and a surface level grid of 80 x 80 points at 25 nautical mile spacing. The analyses were received for every 6 hour period and the forecast periods were 6, 24 and 36 hours from 0000 and 1200 UTC. In general, there was/is insufficient observational data to support independent wind flow analyses for this region using the standard ARAC approach[4] based on observation data.

After ARAC completed the implementation of the data and modeling system for the 3200 km regional calculations, a northern hemisphere modeling system was activated using 381 km gridded analysis data at standard pressure levels. This is the methodology ARAC developed during the Chernobyl accident (1986) for assessment of long range transport and dispersion.[5] Using this modeling system, ARAC evaluated the long range transport, dispersion and deposition of fire generated soot particles at 12-hour intervals (0000 and 1200 UTC).

DESCRIPTION OF MODELING SYSTEM

ARAC's principle atmospheric transport and diffusion models are the Meteorological Data Interpolation Code (MEDIC), the Mass-Adjusted Three-dimensional Wind (MATHEW)[6] model and the Atmospheric Dispersion Particle-In-Cell (ADPIC)[7] model. The MEDIC code was substantially modified to ingest the new RWM data grids along with available observational data to initialize the three-dimensional modeling volume for the regional calculations. The processes involve $1/R^2$ weighted interpolation of the AFGWC data to the ARAC Universal Transverse Mercator (UTM) projection 3200 km grid and vertical interpolation to 15 evenly spaced levels from the surface to 6 km. A detailed regional terrain grid was inserted as the lower boundary of the initialized model volume prior to iterative adjustment by the calculus of variations encompassed in the MATHEW model to achieve a minimally-adjusted, mass-consistent (divergence free) wind field. This process was repeated for every analysis and prognostic data set. The most significant consequence of this MEDIC/MATHEW process appears to be the explicit topographic influence of the flow fields leading to enhanced channeling over the Gulf and delineation of vertical shear. ADPIC is a three-dimensional dispersion model that releases and tracks thousands of marker particles in the flow. These particles are then transported with the wind as they diffuse and are affected by size-dependent gravitational settling and dry deposition processes.

Marker particles from four oil field source regions (up to nine sources are possible) were simultaneously injected into the wind flow fields by ADPIC. Each source region had its own release rate, particle size distribution, mean deposition velocity, and plume rise characteristics. A total of (up to) 20,000 marker particles were used to represent all sources. The smoke particles were given a median diameter of 0.3 microns and a deposition velocity of 1.0 cm/sec. For the fires, plume rise is controlled by the amount of heat energy being released, the inversion height (if present), the stability of the atmosphere, and the speed of the wind in the atmospheric boundary layer. We placed a maximum limit of plume rise at 2000 metres, based on early reports by the United Kingdom and U.S. flight programs. The boundary layer depth and stability were specified for each 6-hour period due to the lack of direct observational data. A diurnal cycle of mixing/boundary layer depth ranging from 500 metres during the nights to 1200 metres during the day, and stability class E (stable) at night to class B (moderately unstable) during the day was modeled.

The air concentration of the smoke was calculated in seven horizontal layers in the vertical, and deposition of the particles was computed at the ground. Contours of ground deposition and vertically integrated optical depth were generated in order to delineate the relatively dense smoke plume structure both for the research flight program and for comparison with available satellite visible imagery. The marker smoke particles serve as tracers for the depiction of the (modeled) flow regimes just as the real smoke particles serve the same function for the actual flow as evidenced by the smoke plumes observed in weather satellite imagery.

DEVELOPMENT OF THE WIND FLOW REGIMES

Beginning with the commencement of combat by Coalition forces on January 17, 1991 ARAC archived all available meteorological data including the RWM analyses from AFGWC. In addition to employing these data for several special assessments during the Gulf war, ARAC staff attempted a simulation of an early oil storage area fire at Al Ahmadi. This initial work paved the way for subsequent contingency assessments. With the quick ending to the combat phase on February 28, all ARAC activities related to the war ceased except the data archival.

When the WMO called a meeting of experts for late April, ARAC evaluated the possible ways its modeling system could contribute to the scientific effort to collect data on the oil fire smoke plumes. It was determined that ARAC could model, both diagnostically and prognostically, the air flow throughout the Gulf region by a combination of the AFGWC RWM analysis and forecast wind components in conjunction with the terrain-influenced, mass-adjusted MATHEW wind flow model. The combined wind field provided the necessary transport input to the ADPIC pollutant dispersion model.

The terrain data used for these calculations was extracted from an online 10 km worldwide digital database ARAC maintains[8] (acquired from NOAA) known as ETOPO5. NOAA created the source database by merging several sets of the original five-minute resolution data. Figure 1 shows the terrain grid used to represent the topographic features of the region, features which proved very important in modeling the regional flow. A typical coverage of surface level meteorological data is shown in figure 2, clearly showing the large data deficient areas of the region. Figure 3 depicts the gridded data detail available to ARAC interpolated from the RWM model output. ARAC has high confidence in this data source because it incorporates the extensive USAF weather data capture resources, military data and weather-satellite-derived information such as cloud tracked winds and vertical thermal structure (over water). Figure 4 shows the terrain grid used to represent the topographic features on a 381 km northern hemisphere grid which ARAC uses for long range transport and dispersal calculations. Figure 5 depicts an example of the wind data and its spatial patterns typical of this scale.

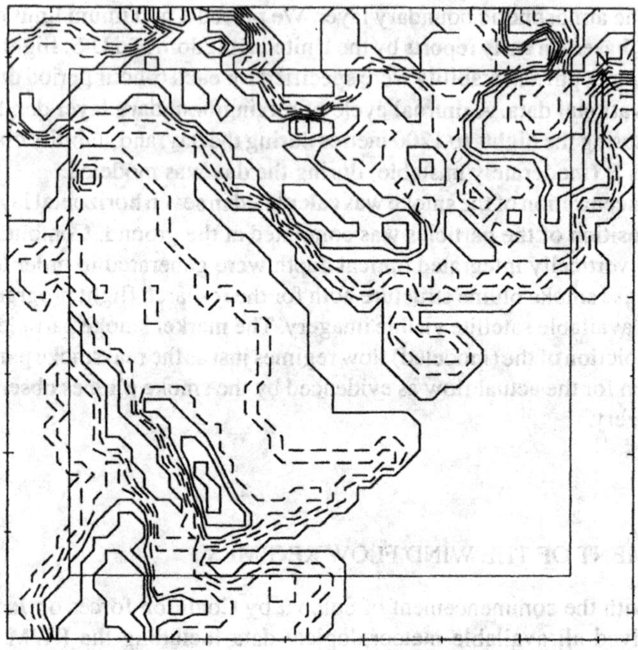

Figure 1. The 50 × 50 terrain grid used in the regional model calculations. Each grid cell is 80 km in the horizontal and the elevation is contoured at ~430 m intervals.

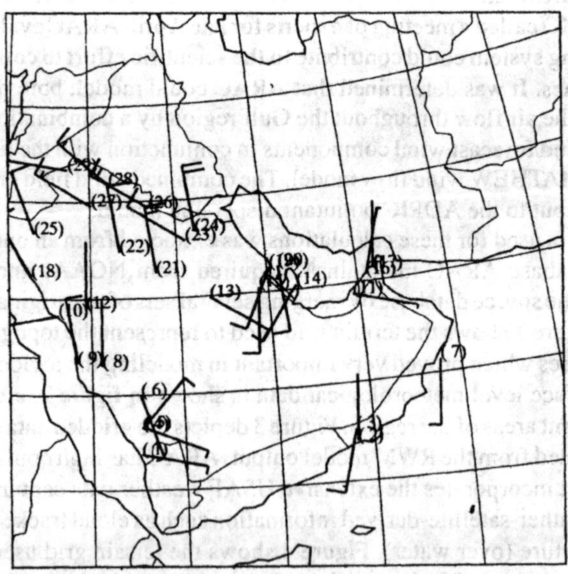

Figure 2. The surface observation data available for a typical primary data period at 0000 UTC, 17 May 1991. Only 29 stations were reported before the meteorological analysis commenced.

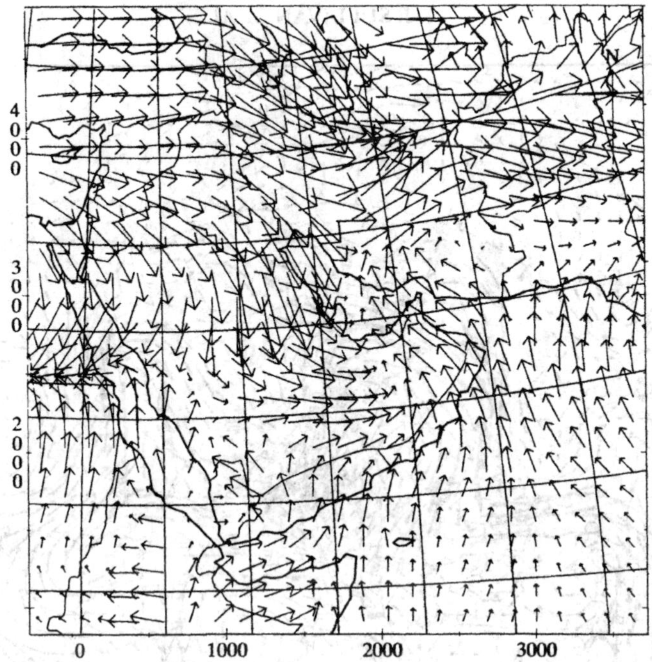

Figure 3. Interpolated wind vectors (for every 2 grid points) at 6 m elevation prepared from the surface observations and AFGWC Relocatable Window Model (RWM) forecast grid. After this data was mass-adjusted, including terrain influence, then the resulting flow field was used for a six hour period centered on 0000 UTC, 17 May 1991.

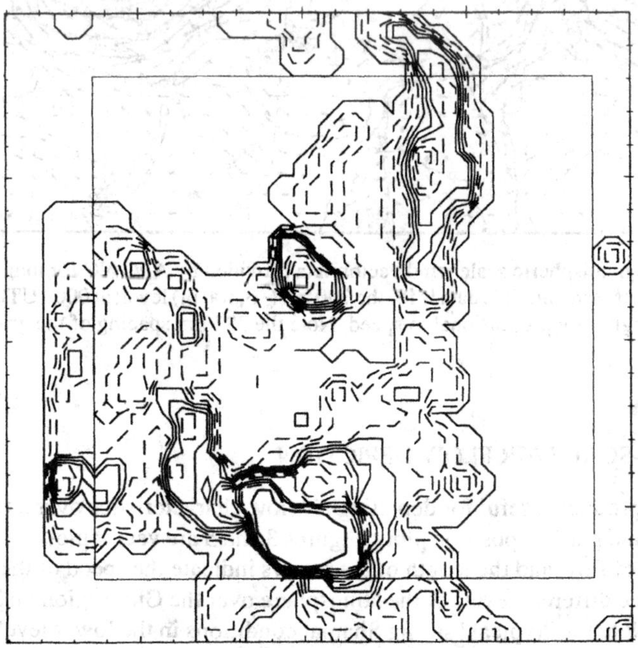

Figure 4. Comparable to Figure 1, this depicts the 47 × 51 381 km terrain grid used with the hemispheric modeling system. The elevation contour interval is 350 m.

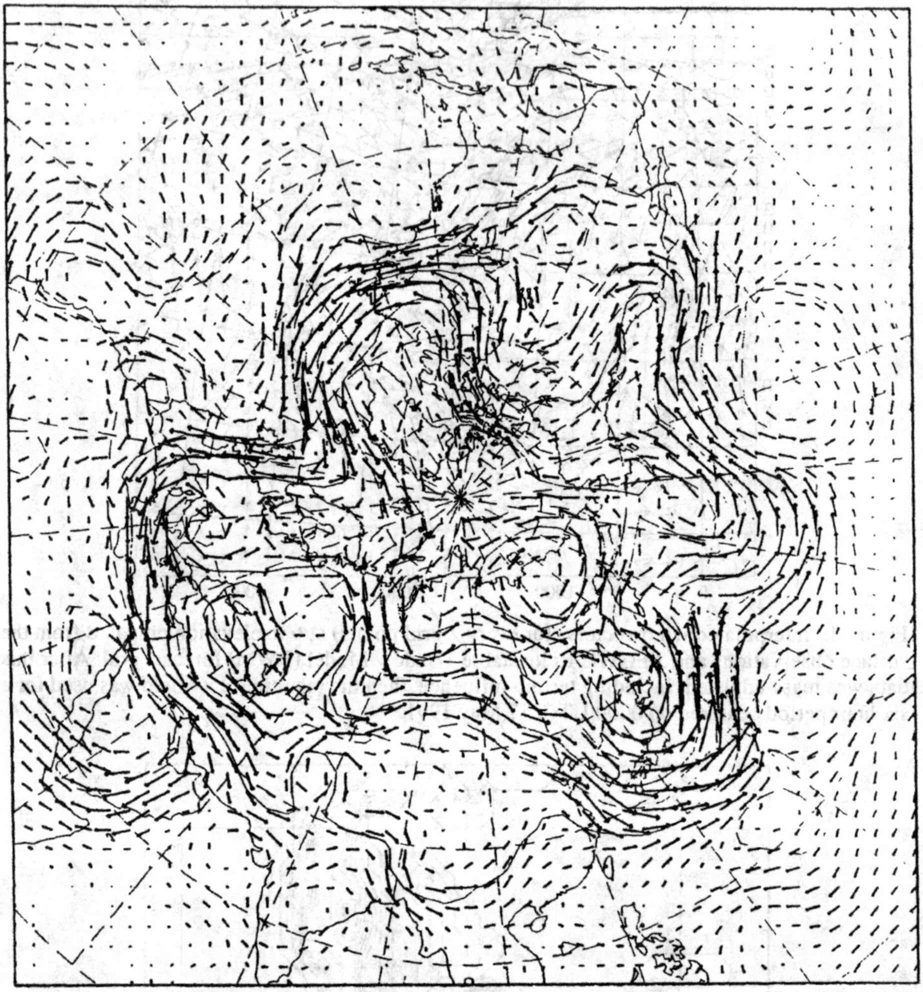

Figure 5. The hemispheric scale wind vector data provided by AFGWC for long range transport and diffusion assessment. This data is for the 700 mb (~3,300 m) level; at 0000 UTC, 17 May 1991. The vector lengths are proportional to speed. Note the 381 km spacing of the grid points.

REGIONAL-SCALE AIR FLOW DEPICTION

Two ARAC products useful for depicting air flow characteristics over a region are wind vector plots and particle position plots. Figures 3 and 5 are vector plots. The arrows show the direction of flow and the length of the arrows indicate the speed of the flow. Figure 6 a–c show three different levels of the atmosphere over the Gulf region on June 3, 1991 at 1200 UTC. Note the "typical" strong Shamal conditions in the lower levels directly over the Gulf and eastern Saudi Arabia on this date. Also note the blocking and deflection of this flow by the Zagros mountains along the western and southern parts of Iran. The missing

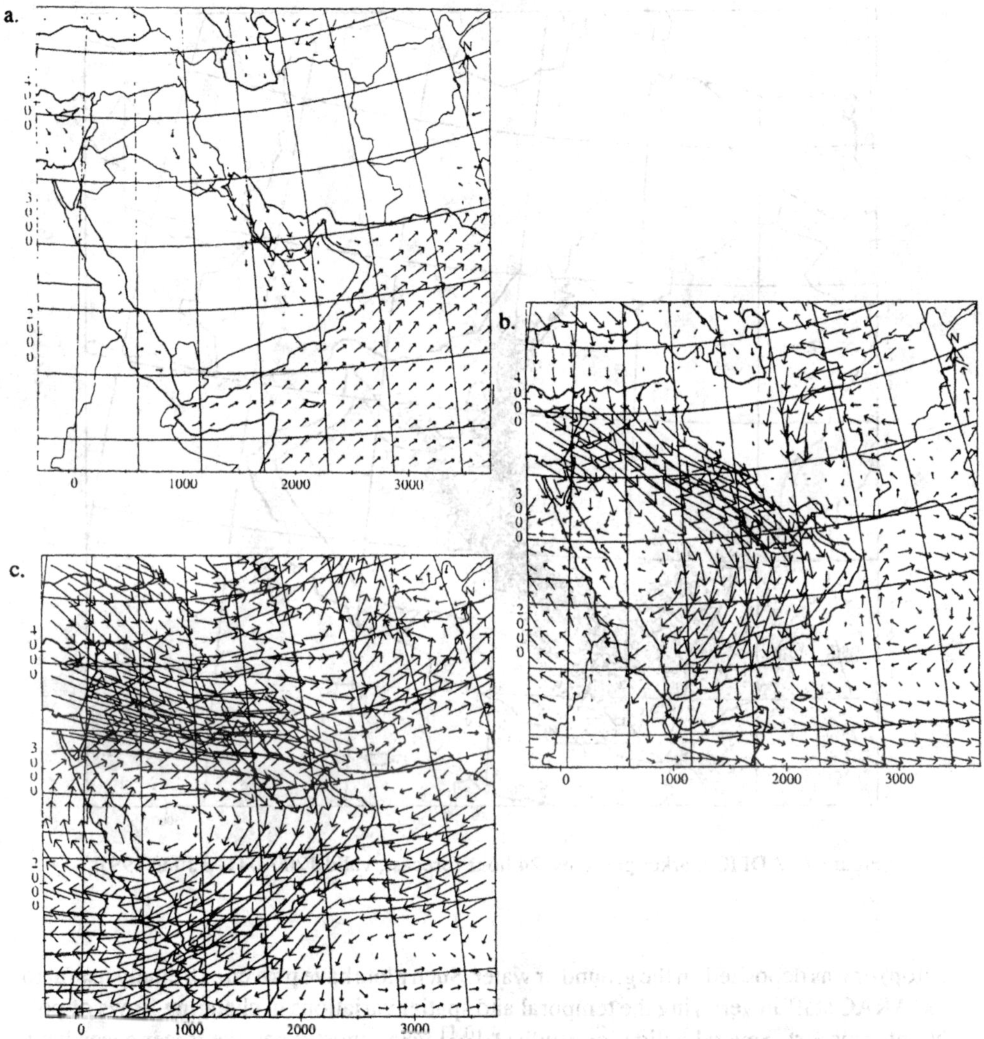

Figures 6a, b, and c. This set shows the mass-adjusted, terrain-influenced MATHEW model output flow fields for a) the surface level, b) the 1714 m level and c) the 3857 m level. Complex, sheared flow environments are evident in several parts of the grid as is the terrain-influence.

and very light winds to the north and east of this boundary indicate the explicit effect of the terrain at the lower levels in this critical region of the model domain. The example wind field also reveals the diverse nature of the air flow over the modeled area on this particular date.

Figure 7 depicts a characteristic particle plot for the forecast of the oil fire smoke plume for June 3, 1991. The plot is an overhead, satellite-like view of the model domain. Since ARAC modeled the fires as continuous releases, the particles depict the resultant flow history for the period of time from when each particle was created until it exited the grid (sides

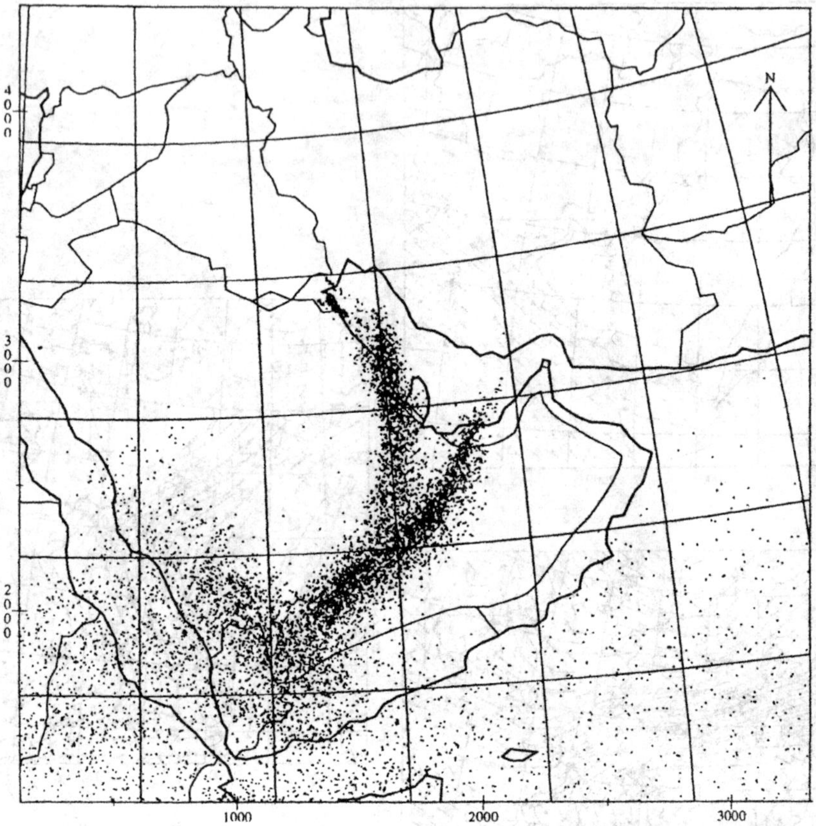

Figure 7. ADPIC marker particles, 24 hour forecast, valid 1200 UTC 3 June 1991.

or top) or was deposited on the ground or water. Such plots have proven extremely useful to the ARAC staff in verifying the temporal and spatial variations in changing flows of pollutant transport. Several validation studies[9,10,11] using mostly passive tracer gases have verified that the particle-in-cell modeling methodology in conjunction with terrain-influenced flow field adjustment is a highly successful technique. However, caution must be used when viewing particle position plots because one cannot directly infer pollutant concentrations from this visual presentation. This is so because the overhead view does not show the vertical location of the particles or the mass of material represented by each individual particle. Particle mass can change due to a time-varying source strengths or because of the numerical computation and particle budget management.

Satellite pictures or imagery of the visible light reflection during daytime can reveal flow patterns if either natural moisture (in the form of clouds) or an absorbing or highly reflective pollutant is in the field of view. In the case of the Kuwait oil fires, weather satellite visible imagery readily detected and revealed the highly absorbent smoke particle plume(s). This has provided a unique opportunity for direct comparison between the actual particles (smoke) and their dispersal by the real atmospheric flow and modeled tracer particles in synthesized model flow regimes derived from limited observational data and mass conservation algorithms. Figure 8 is a copy of the NOAA-11 visible channel image for June 3, 1991 at 1105 UTC. The nearest ARAC particle position plots are for 1200 UTC. The example displayed in Figure 7 shows our 24 hour forecast for June 3rd. Figure 9 depicts the same plume position based completely on a series of analysis data and valid at the same time as Figures 7 and 8. Note that the overall structure of all three depictions is very similar; the plume derived completely from analysis data more closely agrees with the satellite picture in some details and the forecast derived plume agrees equally well in other areas. Overall the essence of the time-integrated flow is well modeled. Figures 10 and 11 similarly compare favorably for May 8, 1991 although it is apparent that the model missed the small, but important differences near the release point. The model maintained the plume immediately on the gulf coast rather than move it a small distance onshore as was detected from the satellite imagery.

Every day was not modeled as well as May 8th. Our forcasts, Figures 12a and b, and the NOAA-11 satellite images, Figure 13, show one day, May 17, when there was a significant problem with the modeled wind flow and subsequent plume position simulation.

Preliminary investigation of the disparity in the calculated versus real plume location points to a lag in detection and analysis of a cold front moving southeastward through the region. Figure 12a is a 36 hour forecast and Figure 12b is a 24 hour forecast. There is some evidence that the latter forecast reflects more of the frontal "push" but it still misses the strong inland penetration of the smoke plume flow over Saudi Arabia. In conjunction with the cold front there was a distinct change in airmass characteristics including a more stable boundary layer and lower and stronger inversion, particularly during the day. An upper air sounding indicated a capping inversion at about 1000 m at 1200 UTC 17 May 1991. Figures 14a, b and c show improved agreement that we were able to achieve by changing our specified diurnal cycle to match the characteristics just discussed. This is another example of the urgent need for reliable and timely meteorological observations which form the basis of regional flow modeling.

For long term continuous plumes, an error in modeled particle positions is propagated until it leaves the model domain. ARAC is attempting to validate its entire modeling effort for the smoke plume transport and dispersal from May 8 to Nov 6, 1991 by means of satellite data intercomparison.

HEMISPHERIC-SCALE AIR FLOW DEPICTION

In addition, ARAC also has begun using hemispheric-scale winds to simulate the long range transport and dispersion of the oil fire plumes. We started with releases beginning on February 21, 1991 and have completed through July 1, 1991 as of this writing. The long range model does not exhibit the details evident in the 3200 km regional model, but the overall patterns based entirely on analyzed wind fields, show strong similarity to the regional results. Thus we are confident that this model represents the plume reasonably well over longer ranges where the smoke particles are too widely dispersed or removed so as not to be visible in the satellite imagery. Figure 15 shows an example of long range plume simulation for May 8 which reveals a very complex and widespread transport and dispersal history for the smoke plume soot particles. Note that since the 1991 version of the ARAC models did not remove smoke particles by cloud/precipitation processes the modeled

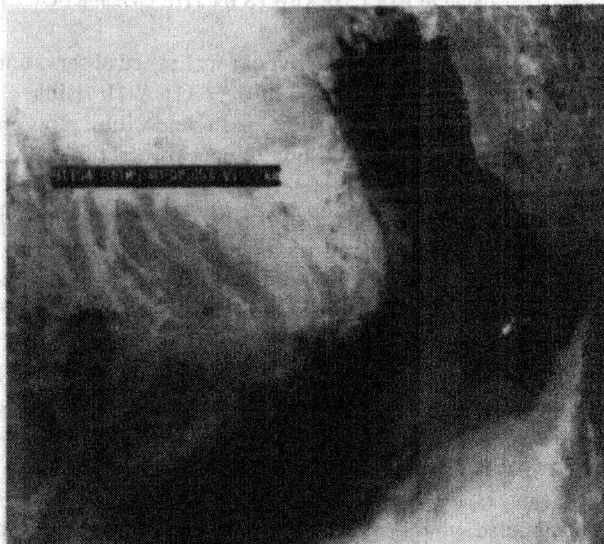

Figure 8. NOAA-11 satellite visible channel image, 1105 UTC 3 June 1991 showing the smoke plume extending deep into Saudi Arabia.

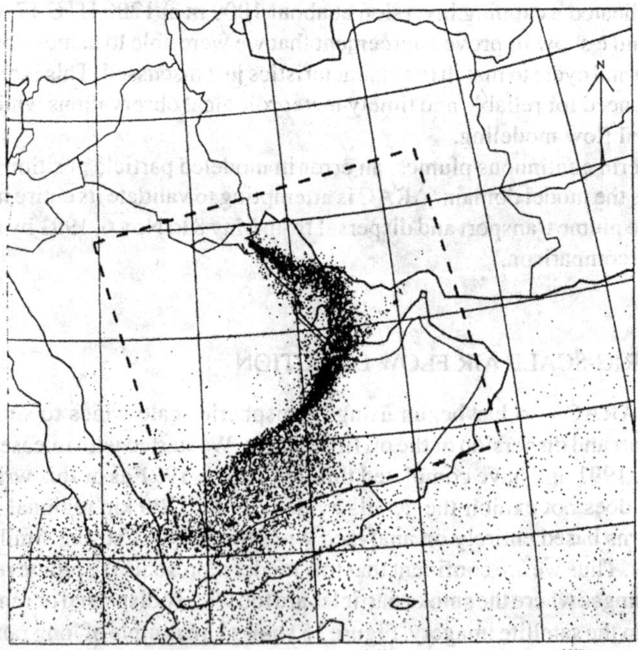

Figure 9. ADPIC marker particles for a refined, analyzed data only plume projection. This shows subtle improvements in the final plume agreement with the satellite image in Figure 8.

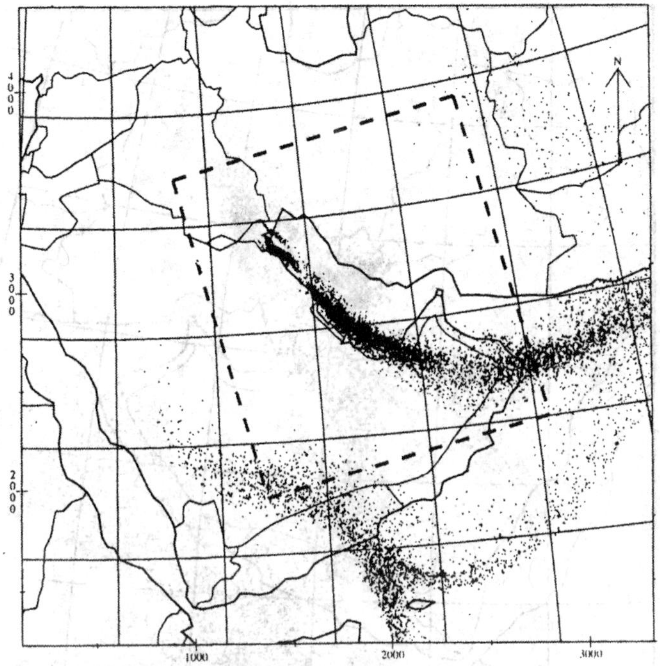

Figure 10. ADPIC marker particles, 0600 UTC 8 May 1991 with a box outlined to match the area covered by the satellite image in Figure 11.

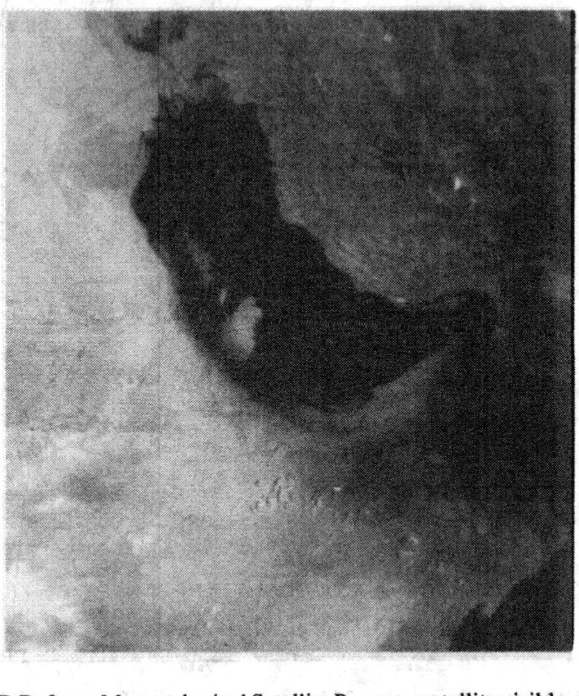

Figure 11. US/DoD Defense Meteorological Satellite Program satellite visible channel image for 0518 UTC 8 May 1991.

a.

b.

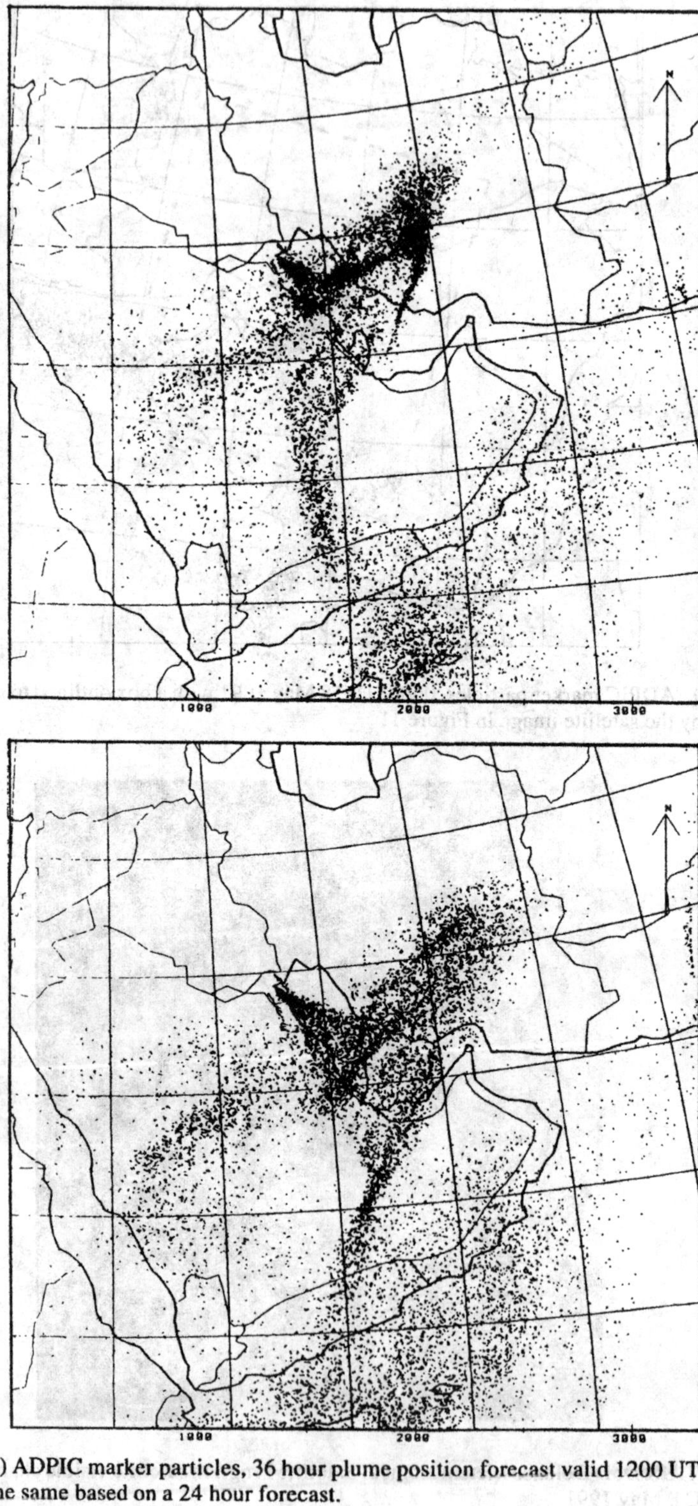

Figure 12. a) ADPIC marker particles, 36 hour plume position forecast valid 1200 UTC 17 May 1991; b) is the same based on a 24 hour forecast.

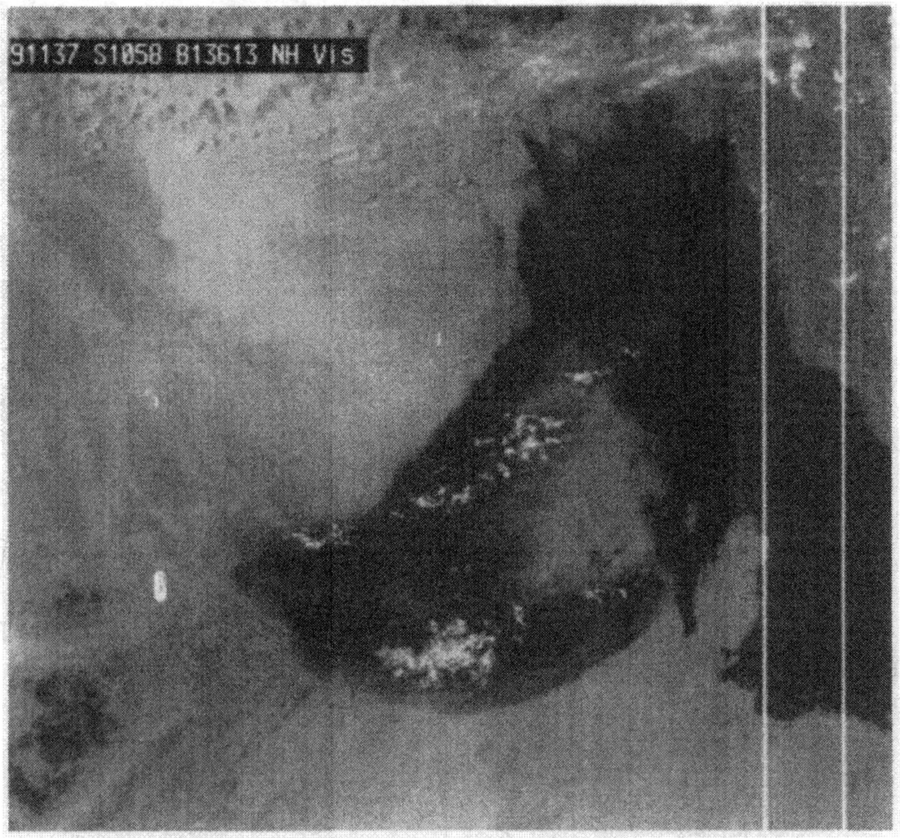

Figure 13. NOAA-11 Satellite visible channel image for 1058 UTC 17 May 1991.

plumes are far more extensive than one would anticipate based on the measured hygroscopic nature of the soot particles.

Some further examples of successful wind flow modeling in the Gulf region are found in Figures 16, 17 and 18 for July 1, 1991. Figure 16 shows our ARAC model particle plot valid for 1200 UTC and Figure 17 is the NOAA-11 satellite visible imagery for just a few hours prior. Note the generally close agreement between these two figures including the thinned smoke veil over Qatar. Figure 18 provides a view of the modeled long range dispersion of the fire soot particles. It is noteworthy to see the basic comparability of these model results and the satellite "truth" data.

Another, and final, example is shown in Figures 19 and 20 depicting the modeled and real plume dispersal for July 25, 1991. Once again there is evidence of substantive modeling skill, both in transport and diffusion. Specifically, the modeling system has very accurately carried the recent plume nearly straight along the western shore of the Gulf and then across the western portion of the United Arab Emirates. A separate, older, sheared segment of the plume is positioned across the southern part of the Gulf and extends well into southern Iran. This provides a very clear example of the three dimensional nature of atmospheric

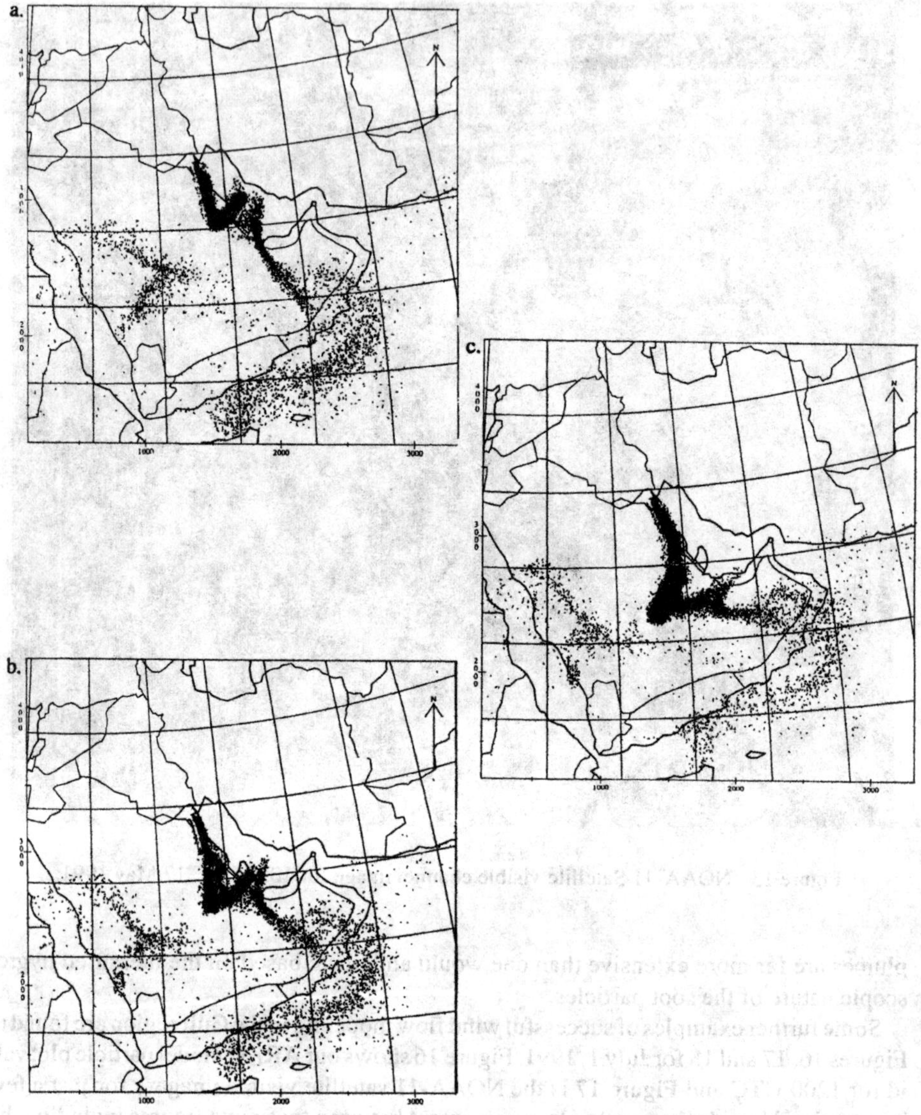

Figure 14a, b, and c. Improved ADPIC plume position and shape, based upon analyzed windflow data and strongly confined mixing layer depth for a) 1200 UTC 17 May 1991, b) 0000 UT 18 May 1991, c) 1200 UTC 18 May 1991.

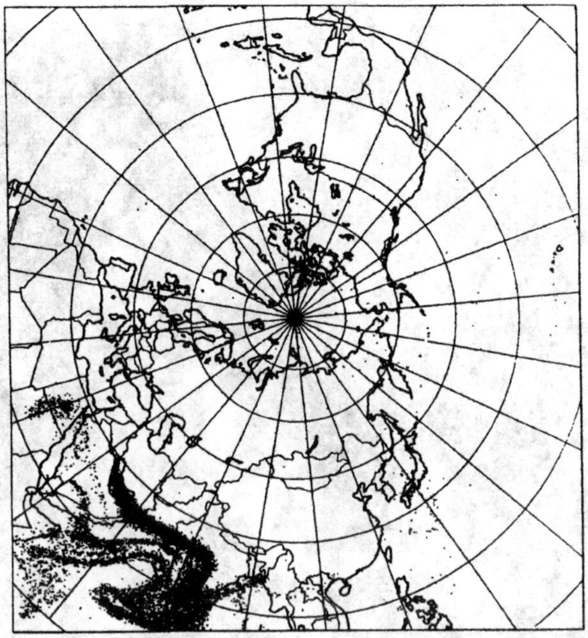

Figure 15. Hemispheric scale ADPIC marker particles valid for 1200 UTC 8 May 1991 based on continuous particle release from 21 February 1991.

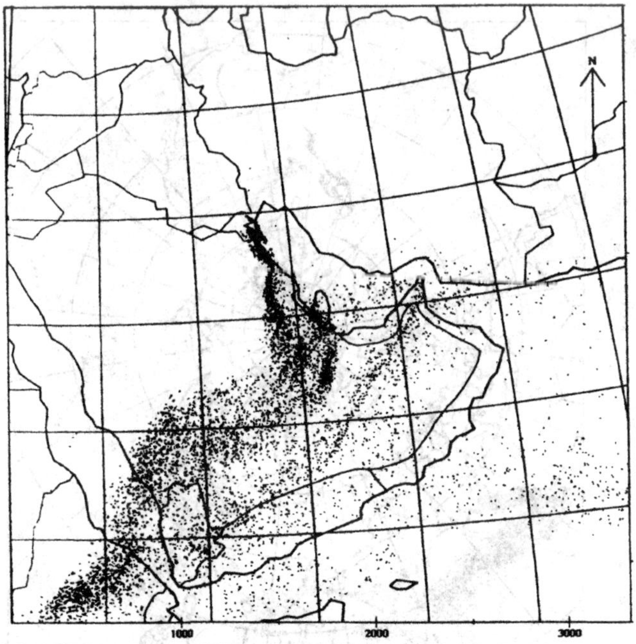

Figure 16. ADPIC marker particles, 1200 UTC 1 July 1991.

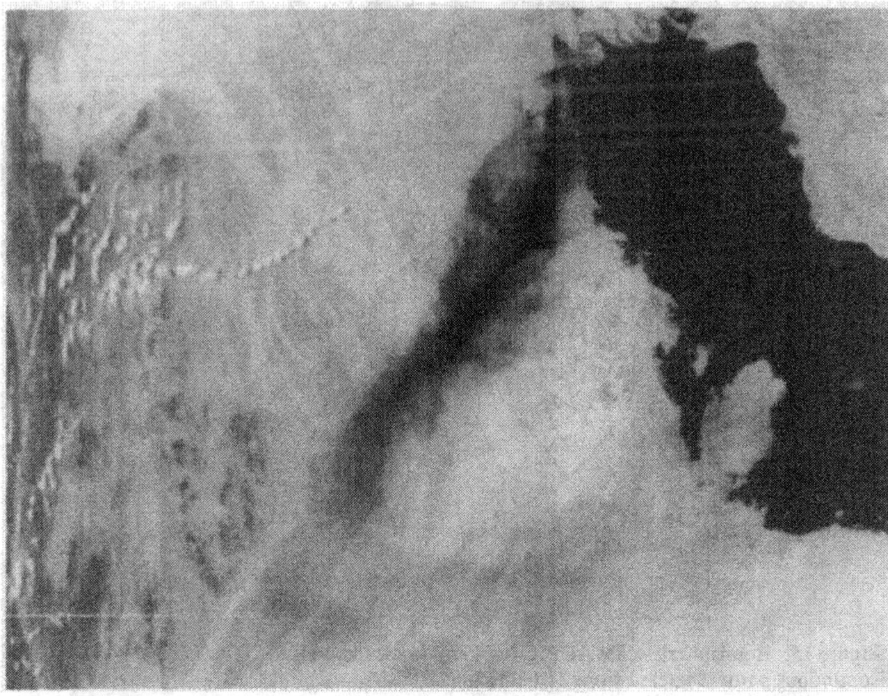

Figure 17. NOAA-11 satellite visible channel image, 1046 UTC 1 July 1991.

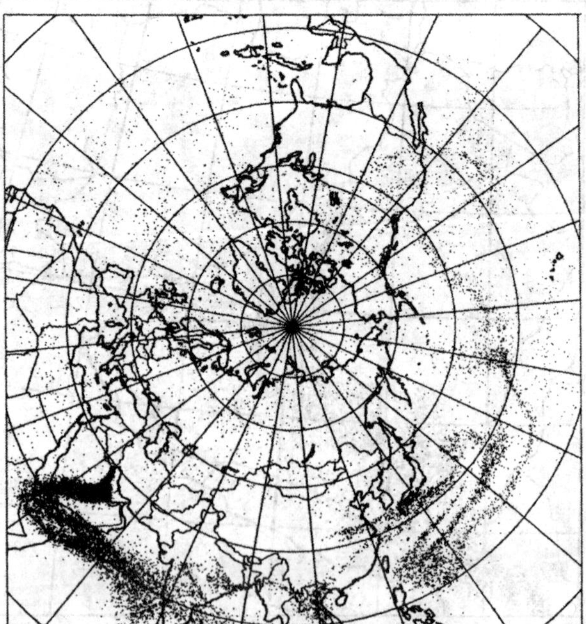

Figure 18. Hemispheric scale ADPIC marker particles, 0000 UTC 1 July 1991.

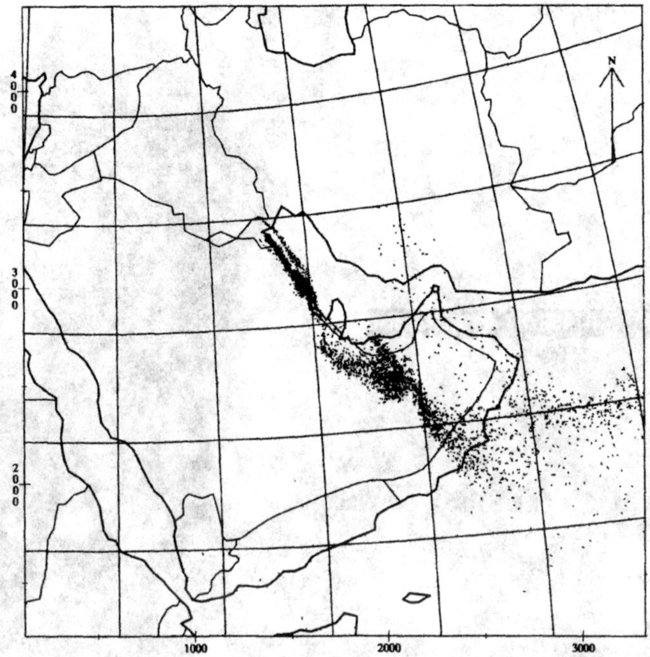

Figure 19. ADPIC marker particles, 1200 UTC 25 July 1991.

flow (in the Gulf region) and the absolute necessity to utilize three-dimensional atmospheric models to assess the transport and dispersion of pollutant emissions throughout the region.

RECOMMENDATIONS

The enormous environmental impacts of the Kuwait oil field fires should be studied and evaluated to the maximum extent possible. Unknown human and ecological consequences should be evaluated and bounded by careful assessment of the pollutant dispersal throughout the Gulf region and beyond. To have confidence in such an approach, the model systems to be used must be evaluated against all available data. The WMO's Background Air Pollution Monitoring Network data must be screened for confirmation (or refutation) of the long range transport of the pollutants out of the Gulf region. Studies should be initiated and supported to determine the human population exposure to hazardous pollutants from the oil fires and also possible impacts or links to precipitation anomalies across South Asia during the period of smoke/soot particle dispersion.

Despite the catastrophic nature of the oil fires, they provided an opportunity to observe and assess the long range widespread dispersion of pollutant emissions. Environmental scientists/scientific teams should be supported in their endeavors to thoroughly research this event with the aim to specifically improve humanity's knowledge of the consequences of inadvertent, accidental and known/planned emissions into the atmosphere.

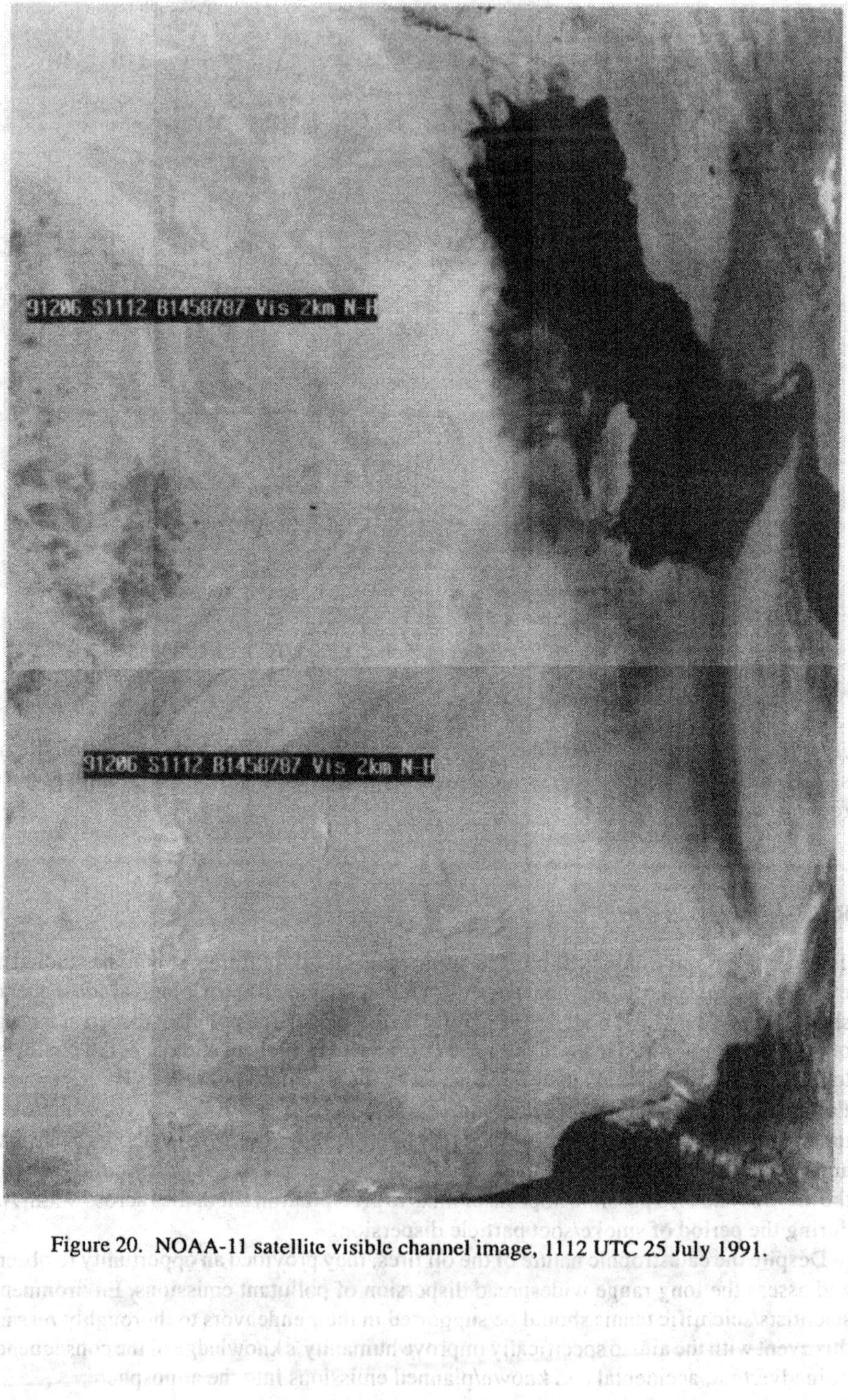

Figure 20. NOAA-11 satellite visible channel image, 1112 UTC 25 July 1991.

SUMMARY

ARAC responded to a U.S. government request to support scientific research flights with forecast Kuwait oil fire plume positions on a daily basis mid-May to mid-June and mid-July to mid-August 1991. Review of the ARAC plume analyses and forecasts by the WMO resulted in an official request from the WMO to the U.S. Department of Energy to have ARAC provide the plume calculations to the Gulf region countries from early June until cessation of the fires.

Both regional and hemispheric modeling systems (and data sources) were employed. A 3200 km regional system, with explicit terrain influenced flow, provided the operational analyses and forecast plume positions. The hemispheric modeling system has been used to study long range aspects of the plume dispersion.

From a limited set of satellite data we see a generally favorable comparison between the ARAC modeled flow and satellite plume imagery. We have found a few occasions with disparate results, most probably a consequence of inadequate input (or availability) of observational weather data to initialize the models. It is strongly recommended that all available data be collected and models be validated and possibly modified to incorporate new knowledge acquired from the study of this major environmental event.

References

1. M.H. Dickerson and R.C. Orphan, "Atmospheric Release Advisory Capability," *Nuclear Safety,* **17**, 281 (1976).
2. R. Lange, "Transferability of a Three-Dimensional Air Quality Model Between Two Different Sites in Complex Terrain," *J. Climate & Applied Meteorology*, **28**, 665–679 (1989).
3. M.H. Dickerson and T.J. Sullivan, "ARAC Response to the Chernobyl Reactor Accident," Lawrence Livermore National Laboratory Report, Livermore, CA, UCID-20834 (1986).
4. T.J. Sullivan and S.S. Taylor, "A Computerized Radiological Emergency Response and Assessment System," *Proceedings of an International Symposium on Emergency Planning and Preparedness for Nuclear Facilities*, International Atomic Energy Agency, Rome, Italy, November 4–8, 1985, Report No. IAEA-SM-280/57).
5. P.H. Gudiksen, T.J. Sullivan, and T.F. Harvey, "The Current Status of ARAC and Its Application to the Chernobyl Event", Lawrence Livermore National Laboratory Report , Livermore, CA, UCRL-95562.
6. C.S. Sherman, "A Mass-Consistent Model for Wind Fields over Complex Terrain," *J. Appl. Meteor.*, **17**,, 312 (1978).
7. R. Lange, "A Three-Dimensional Particle-in-Cell Model for the Dispersal of Atmospheric Pollutants and Its Comparison to Regional Tracer Studies," *J. Appl. Meteor.*, **17**, 320 (1978).
8. H. Walker, "Spatial Data Requirements for Emergency Response," Lawrence Livermore National Laboratory Report, Livermore, CA, UCRL-91263 (1984).
9. M.H. Dickerson, "Summary of MATHEW/ADPIC Model Evaluation Studies," Lawrence Livermore National Laboratory Report, Livermore, CA, UCRL-92319 (1985).
10. K.T. Foster and M.H. Dickerson, "An Updated Summary of MATHEW/ADPIC Model Evaluation Studies," Lawrence Livermore National Laboratory Report, Livermore, CA, UCRL-JC-104134 (1990).
11. R.L. Baskett, J.S. Nasstrom, and R. Lange, "The Value of On-Site SODARS Versus Nearest Radiosonde Soundings in Regional Emergency Response Modeling," Lawrence Livermore National Laboratory Report, Livermore, CA, UCRL-JC-103433 (1990).

SIMULATION OF SHORT TERM ATMOSPHERIC DISPERSION OF SO_2 RESULTING FROM THE KUWAIT OIL FIRES

Dr. DHARI AL-AJMI

Environmental and Earth Sciences Division
Kuwait Institute for Scientific Research

Gaussian plume models have proved to be reasonably successful in describing observed concentration patterns. A Gaussian plume model, RAM, has been applied frequently in air pollution studies. This RAM model was used to simulate the dispersion of sulphur dioxide from the Kuwait oil well fires. The results are analyzed and assessed using the specific features of the model.

INTRODUCTION

RAM is an air quality model based on the Gaussian-plume simplification of the diffusion equation which assumes time independence in the input meteorology and concentrations. The model is primarily used to determine short-term (one-hour to one-day) concentrations from point and area sources. The simulation is done using hourly meteorological data for periods ranging from one hour to one year. A default option is also available which may be used for regulatory applications. Use of this option automatically sets certain parameters to preassigned values for consistency with the United States EPA Guideline for Air Quality Models.

The RAM algorithm is a Gaussian-plume dispersion model that calculates short-term pollutant concentrations from multiple point and/or area sources at a user-specified receptor grid in level or gently rolling terrain where a single wind vector for each hour is a reasonable approximation of the flow over the source area considered. A single mixing height and a single stability class for each hour are assumed representative of the area.

Pollutants considered are relatively non-reactive, such as sulphur dioxide (SO_2) and suspended particulates. Both urban and rural situations can be simulated. In the rural mode, the model uses the Pasquill-Gifford dispersion parameters; in the urban mode, those proposed by Briggs based on the work of Pooler-McElroy are used. Plume rise is calculated following the methods of Briggs and both buoyancy rise and momentum rise are included. For point sources, concentrations are determined using distance crosswind and distance upwind from the receptor to each source. For area sources, the narrow plume simplification of Gifford and Hanna is used with the modification that the area sources are not at ground level, but have an effective height.

Meteorology

Kuwait is generally characterized by a dry, hot climate. Two climatic seasons are recognized, namely: a long hot dusty summer extends from April to October; and a relatively short pleasant winter that starts in November and ends in February. March is transitional between summer and winter. Three main factors define the climate of Kuwait. These are variation in sun inclination, distribution of pressure systems, and source of air masses (Al-Ajmi and Safar, 1987). Significant increase in air temperature during summer is due to high sun inclination angle, which reaches 84° in June, and consequently results in an increase in total amount of radiation. In winter, sun radiation is greatly reduced due to the low angle of sun inclination, which reaches 37° in December.

In summer, Kuwait is affected by the strong pressure gradient between the Mediterranean high pressure system and the monsoon low pressure system which approaches the country from the southeast. This situation is responsible for the prevalence of the strong northwesterly winds that cause sand and dust-storms during the summer. Kuwait is affected by the southern extension of the Siberian high pressure system and the low pressure systems migrating from west to east during winter. This is responsible for the occurrence of cold northwesterly winds.

In June and July, the hottest part of the summer, Kuwait is subjected to a hot and dry northeasterly wind driven by tropical continental air masses that form over northwestern India. In August and September, a moist southeasterly wind caused by tropical maritime air masses originating in the Arabian Sea is usually associated with low pressure and increases in the relative humidity.

The climatic aridity of Kuwait is manifested by extreme diurnal and monthly variations in air temperature. The mean maximum temperature in July is 44.7°C and the mean minimum temperature in January is 7.7°C. Daily temperature ranges of 10°C and 13°C were recorded during winter and summer, respectively. Ground radiation inversion is a common weather phenomena in Kuwait. It occurs on 91% of all the nights and has its maximum at 300m at 60% on the nights. This type of inversion decreases the mixing height of the atmosphere and impedes vertical movement of suspended particles. In summer, the weather is mostly dry. The mean relative humidity is 20%, 21% and 24% in June, July and August respectively. It increases in winter to 60% and 65% in December and January, respectively.

Precipitation in Kuwait is low and normally limited to the cool season from October to May. The mean total rainfall is 115mm/year. However, the total annual amount of rainfall fluctuates greatly from year to year and from place to place. It has ranged from 25mm in 1964 to 260mm in 1976. Evaporation is very high, the daily mean is 16.6mm. It reaches 31.0mm/day in summer (July) and decreases to 5.2mm/day in winter (January).

Dust and sandstorms, which are locally known as "Toze", are very peculiar weather phenomena in Kuwait and surrounding areas. Dust storms usually occur throughout the year, but they are more frequent in spring and mid-summer (March to August). Their frequency distribution during this period is 75% of the total number of dust hours during the year. June and July have the highest frequency of dust-storm occurrence, about 50% of the total number of dust-storm hours.

Table 1. Average Percent Frequency of Wind Direction (Kuwait International Airport, 1962–1989)

Months	N	NE	E	SE	S	SW	W	NW	Calm
January	9.2	2.0	7.7	9.9	11.3	2.1	16.5	22.1	19.2
February	11.3	2.4	9.3	11.8	12.2	1.9	13.1	22.4	15.6
March	13.7	3.3	11.6	12.0	13.7	2.2	10.4	19.4	13.7
April	15.2	3.7	13.5	12.0	15.9	3.1	7.8	15.2	13.6
May	20.9	5.2	13.5	7.6	10.8	2.8	8.2	19.1	11.9
June	19.2	2.3	5.1	2.0	4.9	1.8	14.5	42.7	7.5
July	11.5	1.8	6.1	2.4	5.8	2.1	20.1	41.5	8.7
August	12.1	1.7	7.0	3.7	8.0	2.4	20.3	34.3	10.5
September	13.7	2.5	9.6	4.6	11.8	3.1	11.5	26.1	17.1
October	11.2	2.8	12.3	8.3	16.7	3.3	9.7	17.5	18.2
November	13.4	3.0	8.7	8.0	11.7	2.4	13.9	20.4	18.5
December	9.4	1.6	6.5	8.7	11.2	2.0	17.3	23.6	19.7
Total	11.8	2.7	9.3	7.6	11.2	2.4	13.6	25.4	16.0

Wind direction

Winds blow mainly from two directions, the northwest and to a lesser extent the southeast. Winds from other directions are less frequent and of shorter duration (Table 1). Northwesterly winds are more frequent during summer when it forms 42.7% and 41.5% of the total wind in June and July, respectively. In the spring (March, April and May) the wind direction pattern changes when the frequency of south and south-easterly winds relatively increases (Fig. 1).

Wind Speed

Statistical analysis of wind data collected during the last 27 years (1962–1989) indicates that the average wind speed in Kuwait is about 4.8m/s; however, 16% of the year has a calm weather. Wind speed significantly varies with direction and season (Table 2). It approaches its maximum velocity during summer and is reduced during winter. Average wind speed is 6.2m/s and 4.1m/s during June and January, respectively.

With respect to directions, northwesterly and northerly winds have the highest average wind speed, 6.45m/s and 5.4m/s, respectively. They are followed with the southeasterly and easterly winds which have an average speed of 5.0 m/s and 4.6m/s, respectively. Figure 2 shows the monthly variations in wind speed for the most prevailing directions. It can be noticed that on a monthly average the northwesterly and northerly winds are the most powerful during summer (May–September), while the southeasterly and southerly winds (Fig. 3) are active during winter (November–April).

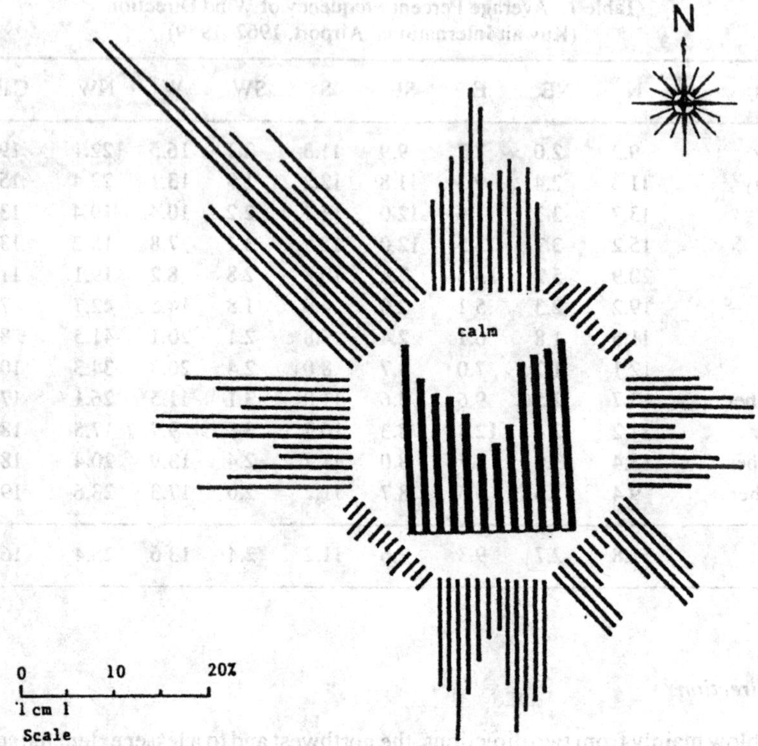

Figure 1. Average monthly frequency % of wind from 8 directions. Months are arranged in clockwise direction, starting with January.

INPUTS REQUIRED

Inputs to the model are a set of options selected by the user, source parameters, meteorological data, and receptor information. Using hourly meteorological data, SO_2 concentrations are calculated for receptor locations either specified by the user or generated by the programme. Emissions and source parameters for point or area sources are required inputs. The meteorological data base, and hence the simulation, can vary from one hour to one year. Concentrations for five averaging periods can be computed. For long-term runs such as a year, a high-five tabulation can be obtained to determine the highest and second highest concentrations at each receptor for each of five averaging periods. If the receptors are specified by the user, their names and coordinates must be provided. RAM requires meteorological and emission data, and depending on receptor options used, it may also require receptor data.

Table 2. Average Monthly Wind Speed (m/s) for Eight Directions (Kuwait International Airport, 1962–1989).

Months	N	NE	E	SE	S	SW	W	NW	Total
January	4.2	3.0	4.6	5.7	4.3	3.2	3.6	5.0	4.1
February	4.6	3.2	4.7	6.4	4.2	2.8	3.6	5.3	4.7
March	5.3	3.9	5.2	6.3	4.5	3.4	4.1	5.7	5.1
April	5.4	3.7	5.5	5.8	4.2	3.8	4.1	5.6	5.0
May	6.1	4.5	4.1	5.5	3.4	3.6	3.6	5.7	4.9
June	7.0	3.5	4.8	4.2	2.7	2.3	4.5	7.4	6.2
July	6.7	3.9	4.9	3.9	2.6	2.1	4.3	7.1	5.7
August	6.1	3.8	4.7	4.3	2.9	2.3	4.3	7.0	5.4
September	5.3	3.0	4.3	3.8	2.6	2.2	2.9	5.7	4.3
October	4.7	3.8	4.6	4.6	3.1	2.6	3.5	5.2	4.2
November	4.7	4.0	4.4	5.2	3.7	2.6	3.0	4.8	4.2
December	4.3	3.0	4.0	6.1	4.1	2.5	3.2	5.0	4.3

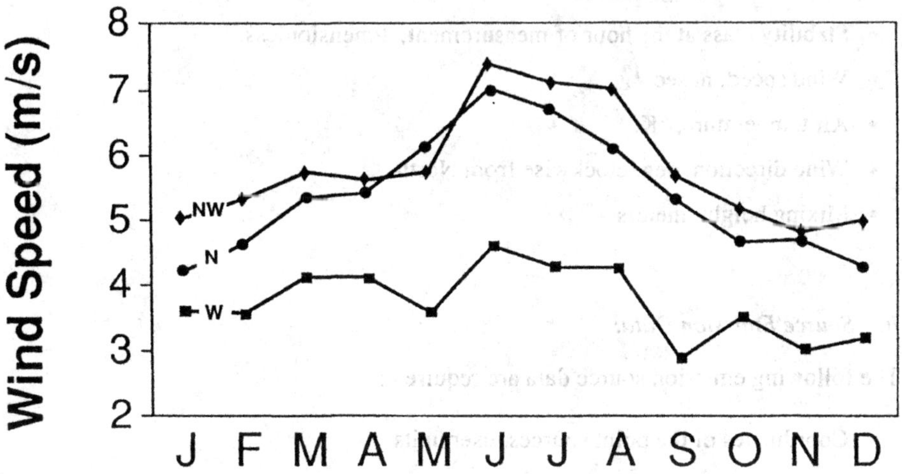

Figure 2. Average monthly wind speed for NW, N and W.

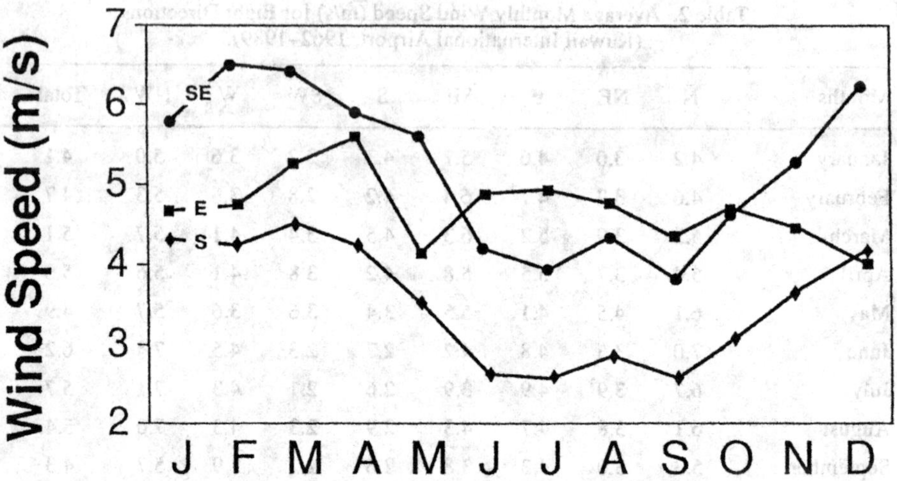

Figure 3. Average monthly wind speed for SE, E and S.

A. Meteorological Data:

The meteorological data required for the model are :

- Power-law wind profile exponents for each stability class.
- Anemometer height, meters.
- Stability class at the hour of measurement, dimensionless.
- Wind speed, m sec^{-1}.
- Air temperature, °K.
- Wind direction, deg. clockwise from North.
- Mixing height, meters.

B. Source Emission Data:

The following emission source data are required :

- Coordinates of the point sources, user units.
- Pollutant emission rate, g sec^{-1}.
- Physical stack height (above ground), meters.
- Stack gas temperature, °K.
- Stack exit diameter, meters.

- Stack gas exit velocity, m sec^{-1}.
- Coordinates of southwest corner of area source, user units.
- Side length of area source, user units.

C. Receptor Data:

The coordinates of the receptors (up to 180) or one to five radial distances may be input. In the latter case, RAM will generate 36 receptors for each distance entered. If the array boundaries are specified, RAM can generate its own honeycomb array of receptors. Additionally, RAM can generate receptors downwind of significant point or area sources if the significant source option is used.

COMPARISON BETWEEN RAM AND SCREEN MODELS

Two models; RAM and SCREEN (EPA, 1988) were applied using the same estimated source emission rate of SO_2 for all the burning oil wells together, for 3 m/s wind and neutral stability. The model runs showed that the models computed similar plume-heights with high accuracy.

The ground level concentrations of the two models were also similar but not as close as were the plume lifts. The most likely reason for this seems to be that the two models use slightly different sigma value functions for the same stability class. This is highly likely although both manuals refer to the same basic report (Turner, 1970). The RAM model states that it uses the Pasquill-Gifford functions as cited in Turner (1970). The SCREEN model only refers to Turner (1970) as a source for determining the stability classes from meteorological data.

In SCREEN the height of the mixing layer is determined by a formula that is valid only for neutral conditions, being proportional to the wind speed alone. This is not as useful, because low wind speed under strongly unstable wind conditions will give rise to very high mixing layer heights, while the same low wind speed under stable conditions will result in very low mixing layer heights. In RAM, the mixing layer height is an input parameter, which necessitates that it be estimated from one of the schemes relating mixing layer height to stability. These schemes can be found in Turner (1970) and also in newer standard books on atmospheric dispersion.

RUNNING THE RAM MODEL

The RAM was applied with a source term equal to the total amount of SO_2, the source term estimated for all the burning oil wells at the two major oil fields, Burgan and Raudhatain.

RAM was used with multiple sources and with a grid of receptor points. The model's options were applied with two sources, corresponding to a northerly and southerly aggregated source, and described the SO_2 sources from the burning oils. The options were found to function well as described in the manuals (EPA, 1987 and Trinity, 1987).

The model created a source inventory which is stored on a map of Kuwait and built into the source coordinates in RAM. For different source and meteorological inputs, one can compute the resulting concentrations in any grid of receptor point across Kuwait.

PLUME LIFT HEIGHT

In the computations with SCREEN and RAM the plume lift was found to be on the order of 250 m, much less than the plume height of 800–2000 m used in Janota (1991). In our computations we followed the recommendations in EPA (1988) and computed the plume height from the specifications of a characteristic oil well fire. Janota (1991) on the other hand combined the volume flow from many sources to yield sources with very large buoyancy fluxes. We believe Janota (1991) was too liberal in combining buoyancy fluxes, and thereby obtained too great a plume lift. This is consistent with many investigators, in the early days of the oil well fires, who believed the smoke would reach far into the stratosphere, thereby having implications on global climate. Fortunately this has now been found to be too pessimistic.

Our result may be somewhat too low, because the SCREEN and RAM models are built to be conservative, and a low plume height is conservative because it gives rise to higher ground concentrations. The difference between our approach and that of Janota (1991) reflects a real physical and unresolved problem for dispersion modelling.

RESULTS AND DISCUSSION

RAM was used to simulate the two dimensional dispersion of SO_2 from the burning oil wells. The wells were grouped and divided into two major areas, namely Burgan and Raudhatain.

The model was applied with a 3-month database. Certain days were chosen for analyzing the dispersion of SO_2 during the months of July, August and September, 1991. The days have been selected based on significant measured SO_2 concentrations reported by the Kuwait Environmental Protection Council (EPC monthly reports of July, August and September, 1991).

Figure 4 shows the locations of the cities of Riqqa, Sabhan, Mansoria, and Jahra (as receptors) on the selected grid for the State of Kuwait.

Figures 5–9 show that during selected days in July, the two areas, Riqqa and Sabhan, were affected by the SO_2 dispersion. Figures 10 and 11 show the dispersion of SO_2 for August 8 and August 13, respectively. Figure 10 shows that the SW wind direction caused the plume to cover the eastern cities on Kuwait's coast. Figure 11 shows that the northerly wind was causing the plume from the Raudhatain oil field to cover Mansoria.

Figures 12–16 show the dispersion of SO_2 during selected days in September 1991. It is very clear that during this month the wind direction was varying greatly. In Fig. 12 the wind direction was from the northwest but on the next day the direction was from the south (Fig. 13) causing the plume to cover most of the cities. Figure 14 shows that a northerly wind caused the northern field plume to cover Mansoria. Figure 15 shows that the wind shifted to be from the south causing the plume to cover Riqqa, Sabhan and Mansoria. The wind was directed from the southeast on day 20 (Fig. 16) resulting in a clear sky over Kuwait's cities.

The results of this assessment using a computer program incorporating the RAM model provides historical data on Kuwait oil fires, and may be useful in assessing risks resulting from this catastrophe.

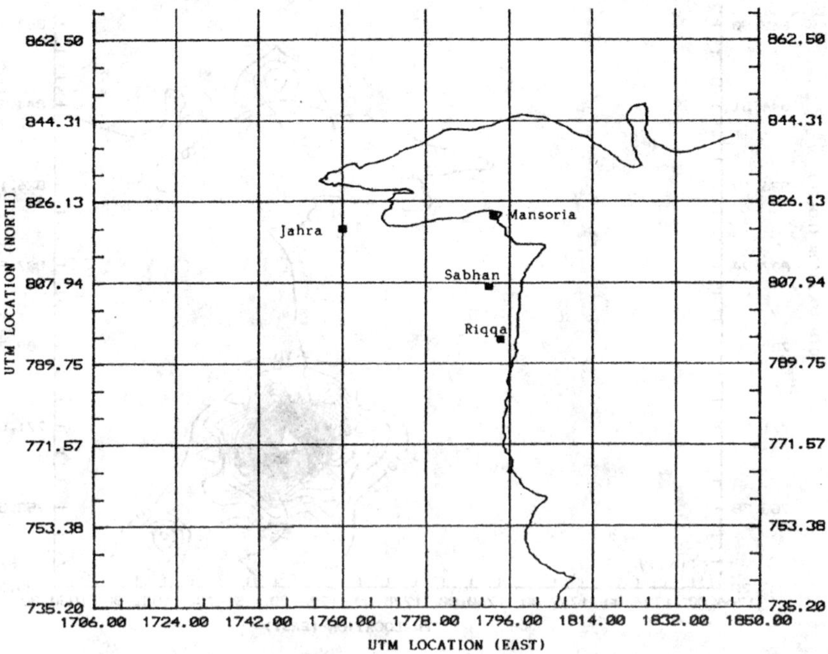

Figure 4. Receptor locations.

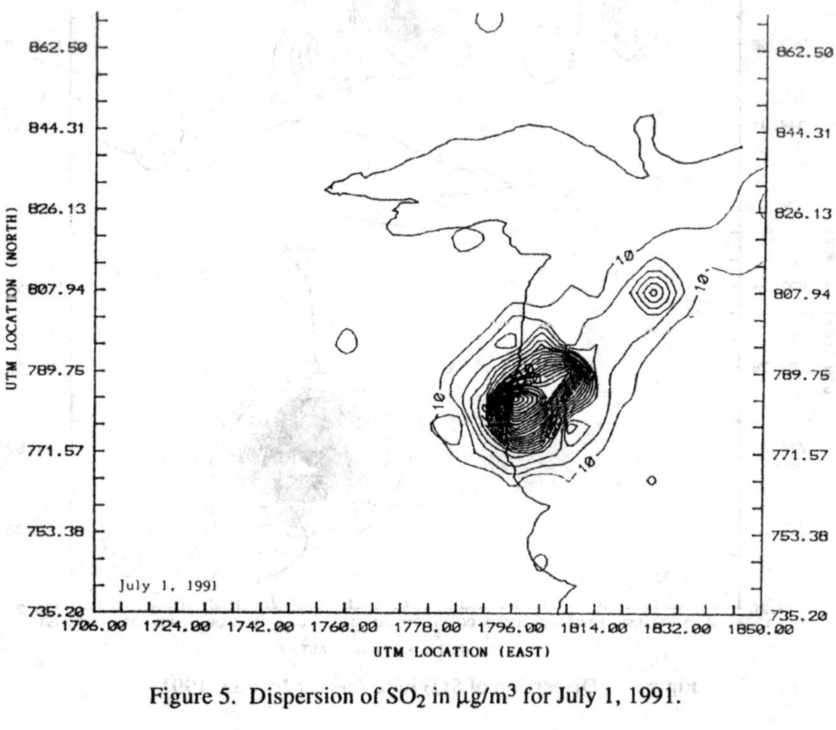

Figure 5. Dispersion of SO_2 in $\mu g/m^3$ for July 1, 1991.

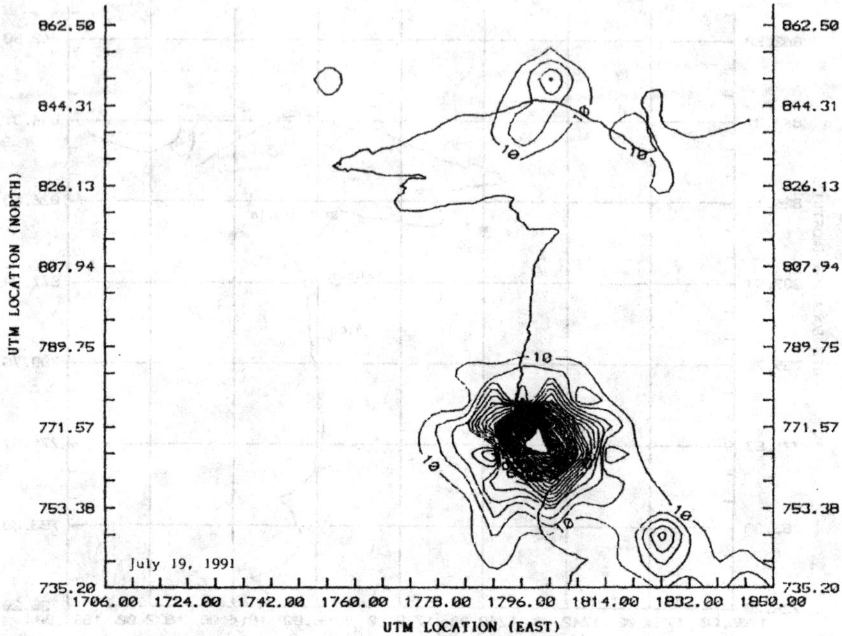

Figure 6. Dispersion of SO_2 in µg/m^3 for July 19, 1991.

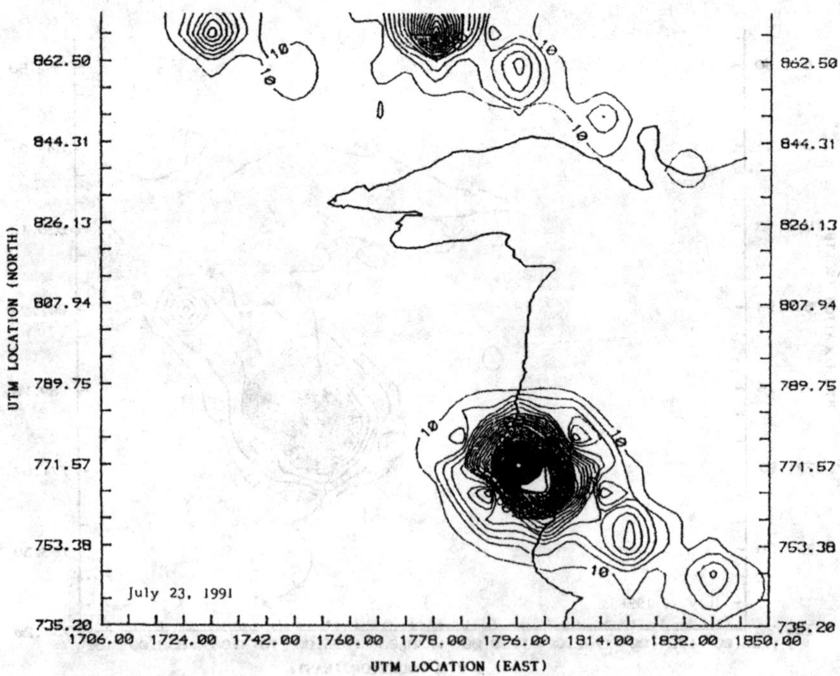

Figure 7. Dispersion of SO_2 in µg/m^3 for July 23, 1991.

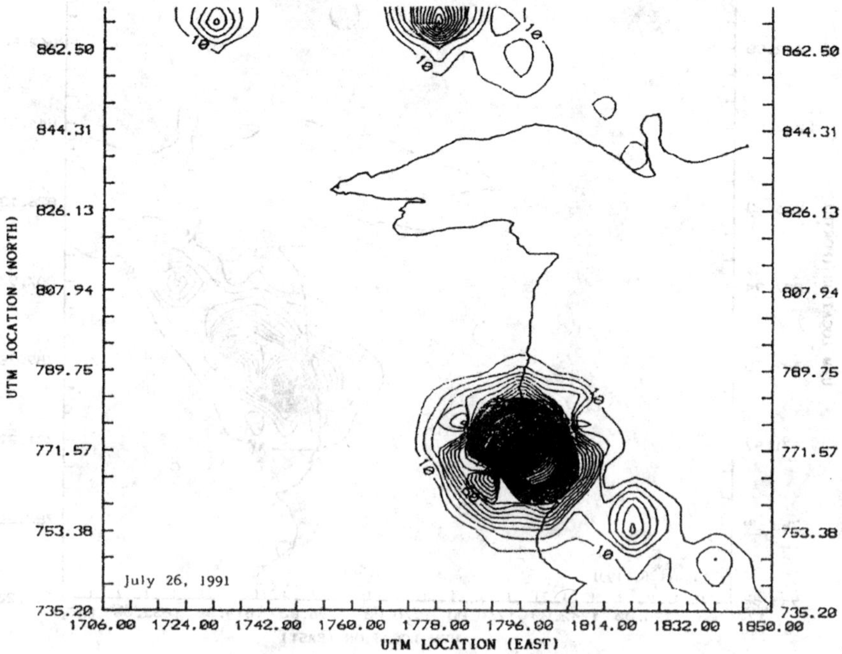

Figure 8. Dispersion of SO_2 in $\mu g/m^3$ for July 26, 1991.

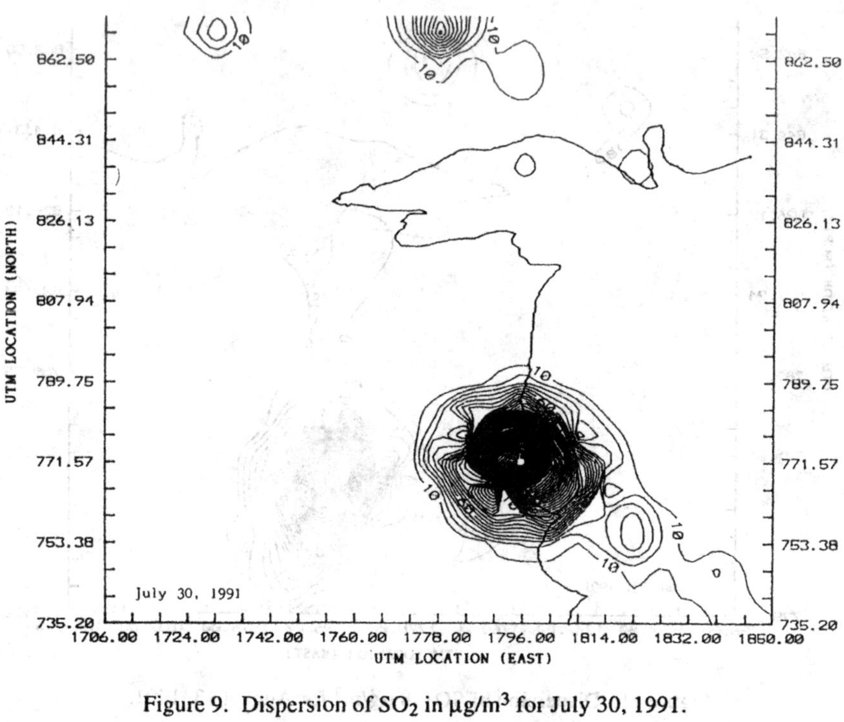

Figure 9. Dispersion of SO_2 in $\mu g/m^3$ for July 30, 1991.

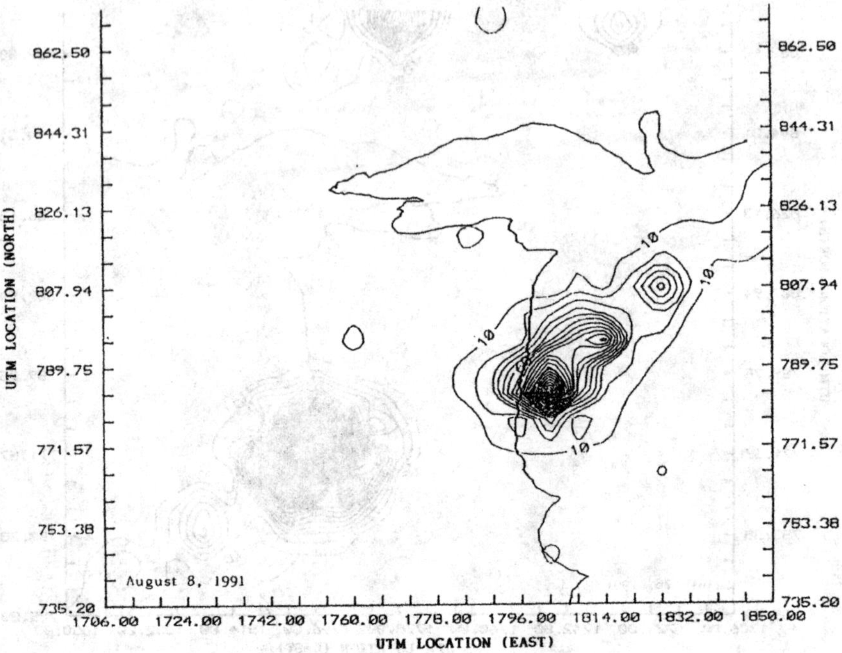

Figure 10. Dispersion of SO_2 in μg/m^3 for August 8, 1991.

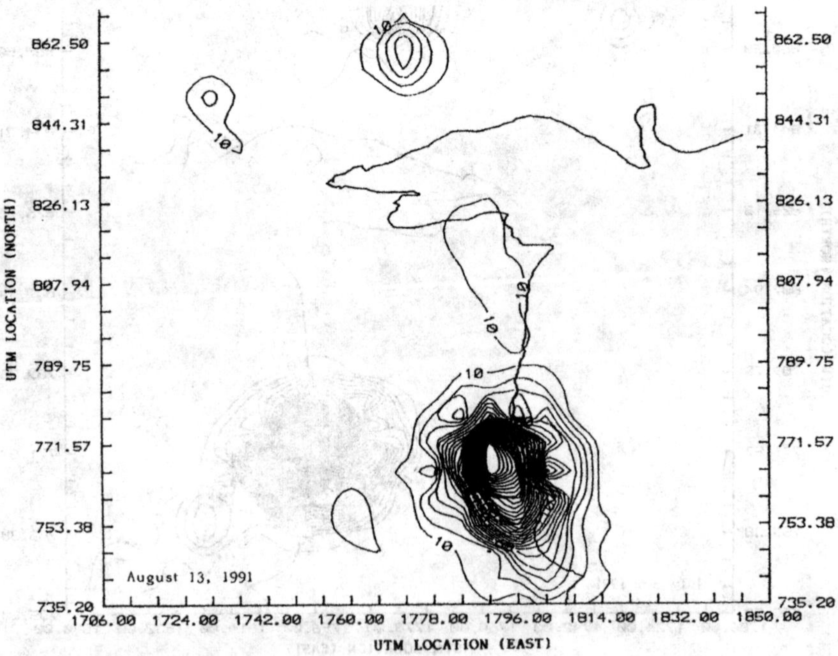

Figure 11. Dispersion of SO_2 in μg/m^3 for August 13, 1991.

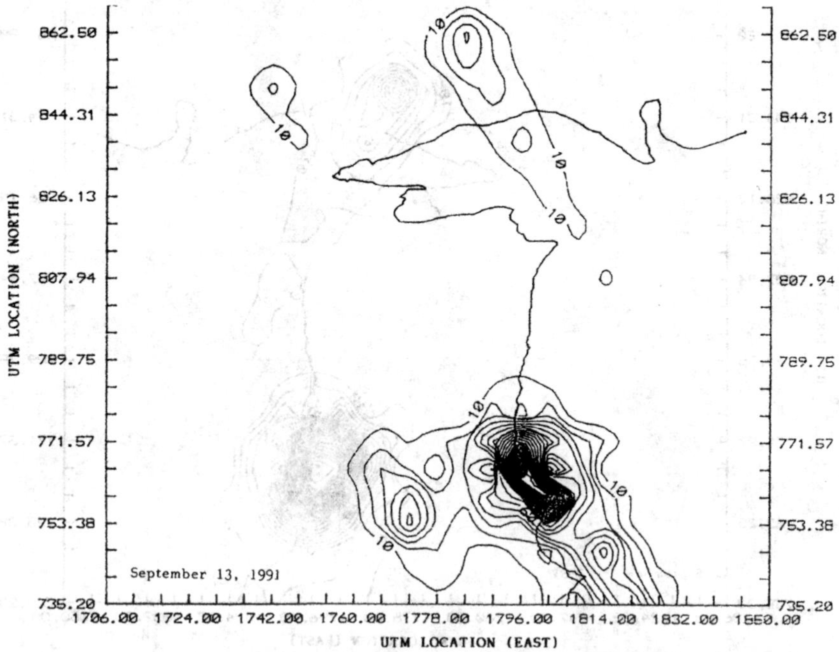

Figure 12. Dispersion of SO_2 in $\mu g/m^3$ for September 13, 1991.

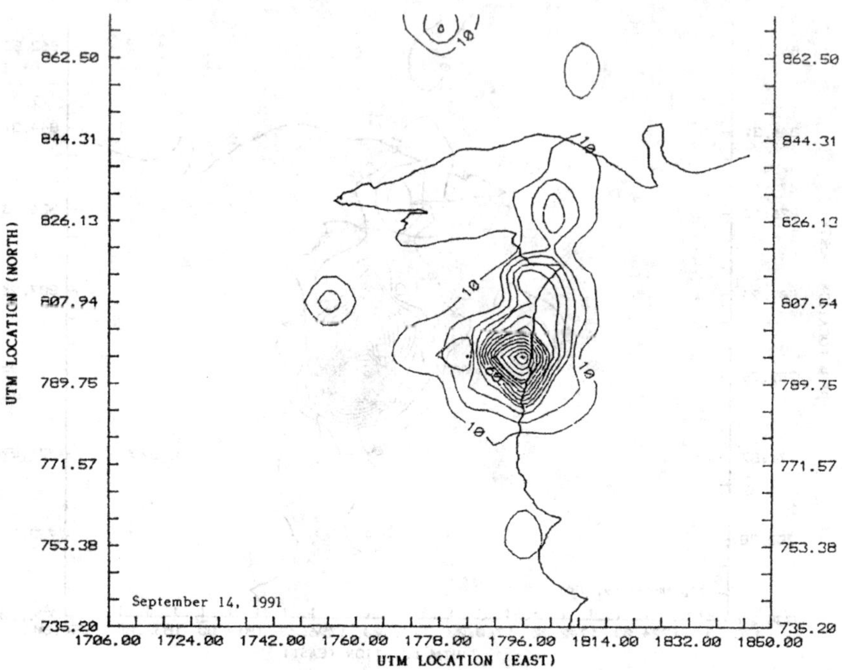

Figure 13. Dispersion of SO_2 in $\mu g/m^3$ for September 14, 1991.

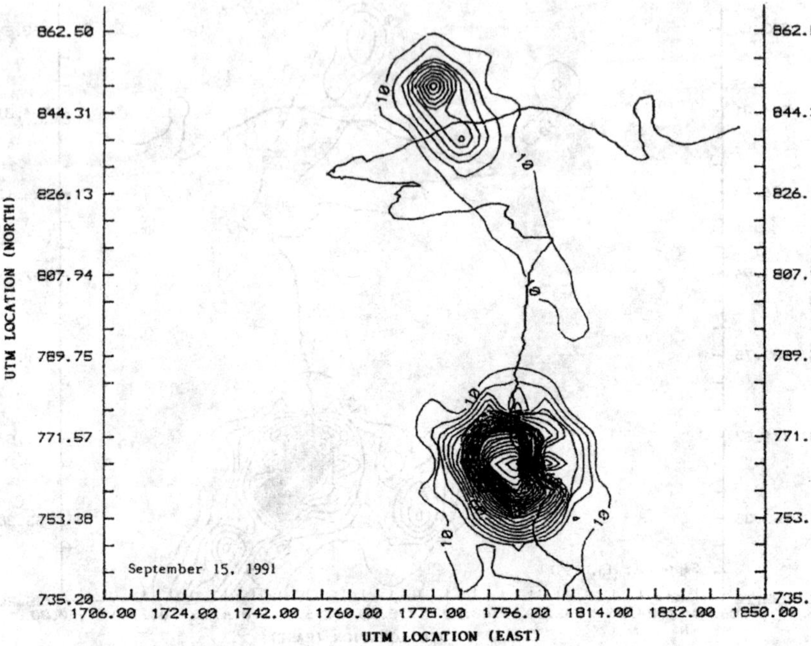

Figure 14. Dispersion of SO_2 in μg/m^3 for September 15, 1991.

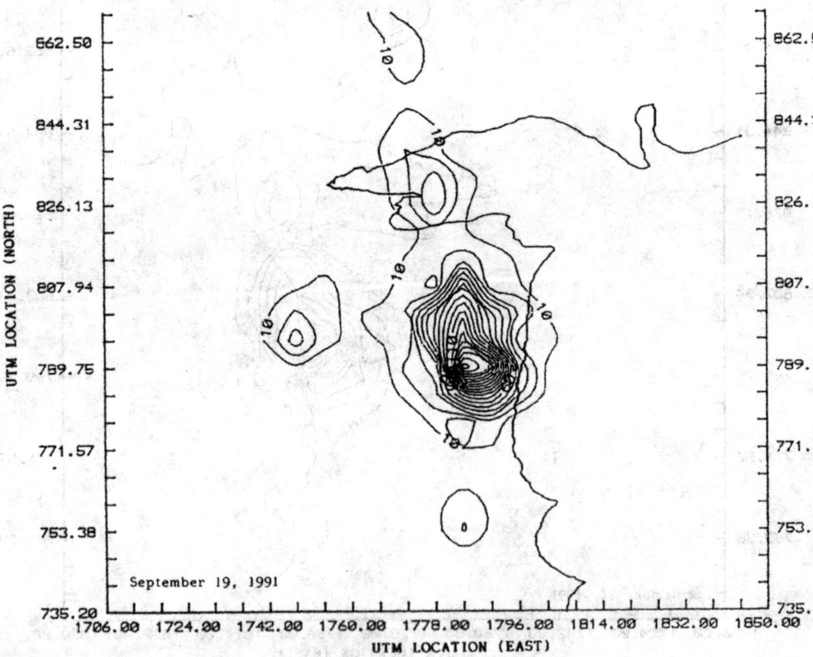

Figure 15. Dispersion of SO_2 in μg/m^3 for September 19, 1991.

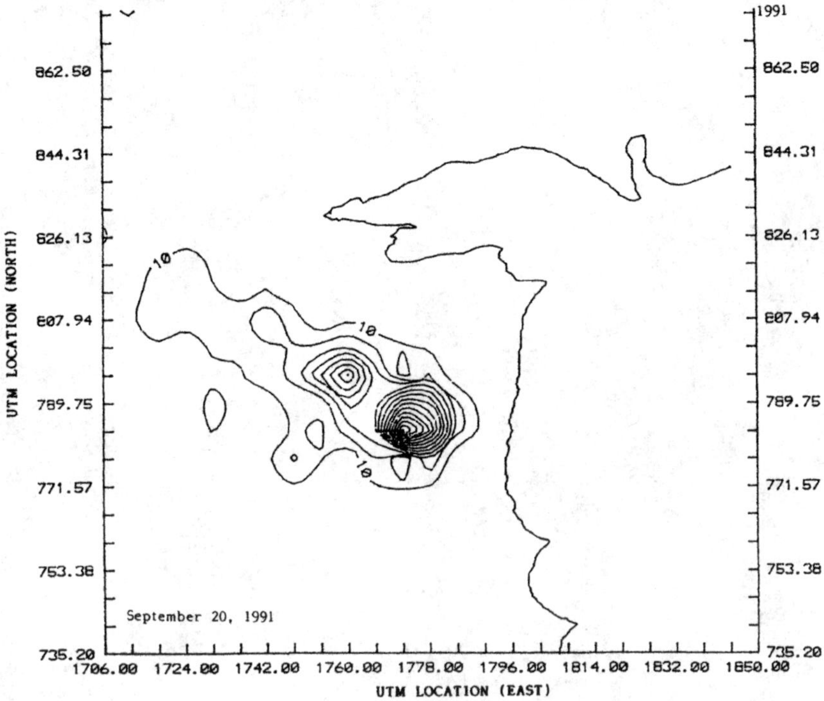

Figure 16. Dispersion of SO_2 in $\mu g/m^3$ for September 20, 1991.

References

Al-Ajmi, D. and Safar, M. 1987. An introduction to climatology and climatic geography. Al-Falah Book Store, Kuwait.

EPA, 1987. Users guide for RAM. (Trinity Consultants).

EPA, 1988. Screening procedures for estimating the air quality impact of stationary sources. (EPA Report EPA–450/4–88–010) including the SCREEN Model Users Guide.

Environmental Protection Council, 1991. Monthly Reports for July, August and September, 1991.

Janota. P. 1991. Air quality damage assessment. (The Analytic Sciences Corporation, 55 Walker Brook Drive, Reading, Mass 01867).

Trinity Consultants Inc. 1987. User manual for RAM. (Trinity Consultants Inc., 12801 N. Central Expwy. Suite 1200, Dallas Texas 75243).

Turner, D.B. 1970. Workbook of atmospheric dispersion estimates. EPA Report no AP–26.

ENVIRONMENTAL DIMENSIONS OF THE GULF WAR: POTENTIAL HEALTH IMPACTS

Dr. FATIMA ABDALI and Dr. SAMI AL-YAKOOB

Environmental Sciences Department
Environmental and Earth Sciences Division
Kuwait Institute for Scientific Research
Address: P.O. Box 24885, 13109 Safat, Kuwait

Iraqis, in the last days before the liberation of Kuwait in 1991, had set on fire 732 oil wells and before that, they have planted nearly one million mines all over the Kuwaiti territories and had spilled nearly 4–6 million barrels of oil in the Gulf which had mainly drifted south to the northern coastline of Saudi Arabia. Accordingly, the year 1992 should be the year of the new environmental perspective, where the human is to be considered as the final and the most important target for any pollutant introduced to the environment.

The Kuwaiti people were exposed to pollution through what they ate, drank, breathed and/or touched during the oil fires. Deleterious impacts on human health and welfare, resulting from contamination of air, water and land resources, must be investigated. Growing concern about these impacts and their immediate, as well as long term, consequences, including risks associated with exposure to emitted oil-related contaminants and the inherent uncertainty of any forecast, makes the prediction and analysis of environmental impacts and risks a task of increasing national and regional importance.

As public awareness of the importance of environmental health grows, the new perspective should define the association between the health of individuals, the pollution of the environment, and the extent of the effect on socio-economic circumstances.

To estimate the risks associated with human exposure to oil pollutants, hazards must be identified, exposures must be characterized and toxicologic and epidemiologic evaluations must be performed. Environmental monitoring results could be used as enviroindicators for the impact assessment studies. However, biomonitoring technology is highly recommended for the exposure and risk assessment programs for following up the short and long-term impacts of the oil-related pollutants on human health.

INTRODUCTION

On January 26th 1991, the Iraqi troops deliberately spilled 4–8 million barrels of oil into the Arabian Gulf. The short-term impact was an oil slick covering a large area of the northern part of Saudi Arabia. On February 21st 1991, the Iraqi army ignited about 732 oil wells in Kuwait, representing one of the most extraordinary man-made environmental disasters in history.

The long-term environmental consequences of the catastrophe are still uncertain and undetermined. Several independent organizations, local and international, have studied the problem utilizing different approaches, however, rehabilitation of the environment is inconsistent. According to several reports, the ecosystem of the Gulf will take at least a decade to recover from the oil spill.

Table 1. Kuwait Oil Field Survey.

		TOTAL WELLS STATUS				
		TOTAL NO. OF WELLS SABOTAGED				
FIELD	DRILLED	ON FIRE	GUSHING	DAMAGED	INTACT	OBS.&ABN.
MAGWA	147	98	6	21	15	7
AHMADI	89	60	2	18	6	3
BURGAN	423	292	24	28	67	12
RAUDHATAIN	83	63	2	5	3	7
SABRIYAH	71	39	4	9	5	14
RATQA	114	1			8	
BAHRA	19	3	2			
MINAGISH	40	27		7	1	
UMM GUDAIR	44	27	3	11	2	2
DHARIF	4	0	0	0	3	
ABDULIYAH	5	0	0	0	4	
KHASHMAN	7	0	0	1	1	5
*SOUTH UMM GUDAIR**	18	0	0	0	16	2
*WAFRA **	482	6	33		15	
SOUTH FUWARIS *	9	0	0	0	9	
TOTAL=	1555	616	76	100	155	52
TOTAL(ONFIRE,GUSHING,AND DAM.)=				792		

BLANK CELL INDICATES THAT WE ARE NOT CERTAIN ABOUT THE NUMBER
* 50% SHARED WELLS WITH SAUDI ARABIA

Source: Kuwait Oil Wells Blow-out, Aspects and Effects, 1991.

Kuwaiti oil is "sour crude," contains high levels of odorous and corrosive sulphur. The inefficient combustion of the uncapped and burning oil well created an environmental catastrophe. The main concern from this atmospheric pollution was the plume of tons of particulates, volatile organic compounds (VOCs), semi-volatile organic compounds and gases of many pollutants such as primary air pollutants: SO_2, NO_x, H_2S; hydrocarbons; sooty smoke; and trace metals which were emitted up to a height of more than 2–3 km. Table (1) shows the Kuwait oil field survey prepared by Kuwait Oil Company (KOC) and Kuwait Institute for Scientific Research (KISR), summarizing the oil fields, and total number of wells sabotaged (Kuwait Oil Wells Blow-out Aspect, and Effects, 1991). Fig. 1 and 2 represent the estimated emissions from the oil fires by the British Team Report (1991) and MEPA Report (1991).

More than 700 Oil Wells have been Set on Fire

The significance of the emmited pollutants on the regional scale

Pollutant (Million tons/Year)	
Black Smoke	5
SO2	2
NOx	0.5
CO	15

Figure 1. Estimated emissions from Kuwait oil fires. Modified from British Report, May 1991.

More than 700 Oil Wells have been set on Fire

It has been estimated that about 2.5 million barrels of crude oil and about 35 million cubic meters of associated gases are on fire.

Emissions	
SO2	20,000 tons
Particulates	1500 tons
CO	250 tons
NOx	500 tons

Figure 2. Estimated emissions from Kuwait oil fires. Modified from MEPA Report, April 1991.

The acute risk coefficient for the population of Kuwait and nearby countries predicted in the Harvard Conference (1991) was significant (excess acute mortality could reach 10% in some areas).

Most of the monitoring studies conducted during the oil fire catastrophe, conducted chemical and physical analysis of contaminants in the air. Although biological monitoring is a sensitive index of an individual's exposure to chemicals, it was less used and studied during the oil fires.

Environmental Monitoring:

To assess the impacts of the Gulf disaster, several perspectives must be reviewed. Many studies found high levels of particulate matter in air samples. Long-term exposure to these contaminants may contribute to chronic and probably to irreversible respiratory disorders. These pollutants, besides their impact on human health, may have adverse effects on the environment.

The U.S. interagency Air Assessment Team has conducted a survey of the fire plumes and found limited concentrations of SO_2 and H_2S near the burning wells (Gulf Regional Air Monitoring Program, 1991). On the other hand, high levels of particulate were found in the air. Accordingly, the susceptible subpopulations, such as hypersensitive individuals with chronic lung disease and asthma, may suffer exacerbation of their symptoms, especially during the worst meteorological conditions such as poor air mixing, or plume touch down.

The U.S. interagency Air Assessment Team detected the highest level of particulates (–5.4 mg/m^3 and VOC's (–2–5 ppm), adjacent to large pools of oil (Gulf Regional Air Monitoring Program, 1991). Particulate level of 480 μγ/m^3 was obtained at MEPA in Dharan, Saudi Arabia. Baseline particulate levels due to dust storm in Kuwait and Saudi Arabia, range from 200–3000 μ/m^3. During March 1991, the highest reading of particulate levels at Al-Ahmadi Hospital, with a 20 min. average reading was 0.935 mg/m^3 (Gulf Regional Air Monitoring Program, 1991). In their results, none of the SO_2 samples collected in populated areas exceeded the EPA air quality limits of (365 μg/m^3 or 0.14 ppm for 24 hr exposure).

Based on their results, very low levels of PAH's and heavy metals were detected in populated areas. The major hazard associated with the oil well fires was the particulate matter. Benzene, ethylbenzene, toluene and xylene and others were among the VOC's detected. Naphthalenes were among the PAH's detected, however, more intensive analysis on PAH's adsorbed on particulates (especially the respirable size of ≤ 10 μm) are needed to investigate long term impacts on health and ecosystem.

Analyses conducted by KISR/EES Division in January-February, 1992, showed significant concentration of some selected toxicants such as those shown in Table 2. One possible source of these pollutants could be the vapours from the remaining oil lakes which are up to 80 to 90 acres in size and estimated to contain 60–100 million barrels of oil scattered throughout Kuwait. Another possible source could be the resuspension by dust storms of the deposited soot which is covering large areas in Kuwait. Most of the studies carried out showed that levels of particulate matter, including the smallest particles that can be present indoors, were much higher than the EPA's maximum allowable standard (EPA's standard for the smallest particles (≤10 μm) allows a maximum of 150 particles/m^3 over a 24-hour period; or 260 μg/m^3 which is not to be exceeded more than once a year).

The WHO (1991) reported that total suspended particulates ranged from 10 μg/m^3 at the US Embassy in Kuwait to 5,400 μg/m^3 in the Al-Maqwa oil field. Al-Yakoob et al. (1992) reported that the high levels of respirable particulates in the air raises a concern about possible deleterious impacts on human health. The long-term exposure to particulate matter may contribute to respiratory discomforts. Figures 3, 4, 5 show the log concentrations of particulate matter in 3 monitoring areas (Al-Mansoriah, Al-Riqqa and Al-Jahra) originally measured by the EPC (1991) during the oil fires. Most of the readings during May-September, 1991, were above the standard level of EPA (260 μg/m^3 for 24-hr exposure). Although Kuwait is known for its highest record of total suspended particulate materials (610 μg/m^3), the main concern is the composition of the fine particulate matter rather than the total weight.

Özkaynak (1991) reported the PM_{10} concentration at 50-830 μg/m^3 with a mean of 328 μg/m^3, and several toxic substances were detected in oil fire smoke. The National Toxic Campaign (A Boston based environmental group) found 1,4-dichlorobenzene, 1,2-dichlor-

Table 2. Mean of Organic Pollutants Detected During January 1992

Location\Pollutant	Benzene	Total Hydrocarbons*	Toluene	Xylene	Nitrobenzene	M-Dichloro Benzene	1,2-Dichloro Ethylene	SO_2
Station (1) Al-Salem								
6:30 A.M.	4.64	0.162	2.176	1.026	0.502	0.061	4.516	0.682
2:30 P.M	1.10	0.535	0.314	0.046	0.024	0.008	0.100	0.072
Station (2) KISR								
6:30 A.M.	0.542	0.168	0.660	3.872	0.350	0.537	20.835	2.627
2:30 P.M.	2.782	0.572	0.638	1.503	0.024	0	0.567	0.536
Station (3) 7th Ring Road								
6:30 A.M.	0.258	0	0.638	5.612	0.608	0.160	3.664	2.530
2:30 P.M.	1.276	C.876	0.838	1.742	0.008	0	0.068	0.154

Modified from (KISR Aurpollution Special Assignment).

*On April 13, 1992 Total Hydrocarbons reading was 2.6 ppm compared to standard value of 0.24 ppm. Annual dose not to be exceeded once a year.

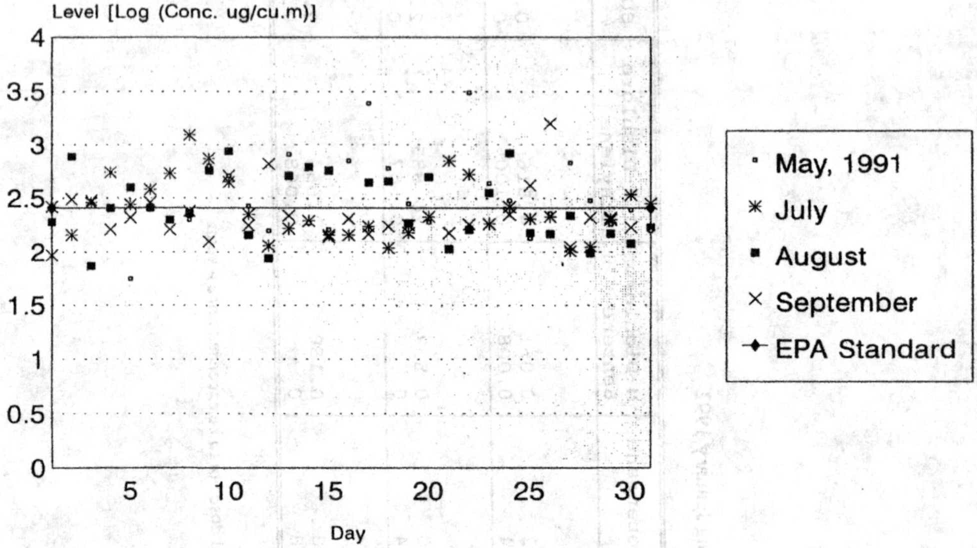

Modified From EPC Final Reports and Al-Majed et. al., 1991

Figure 3. Outdoor levels of suspended particulate matter (< 10 um) Al-Mansourian area

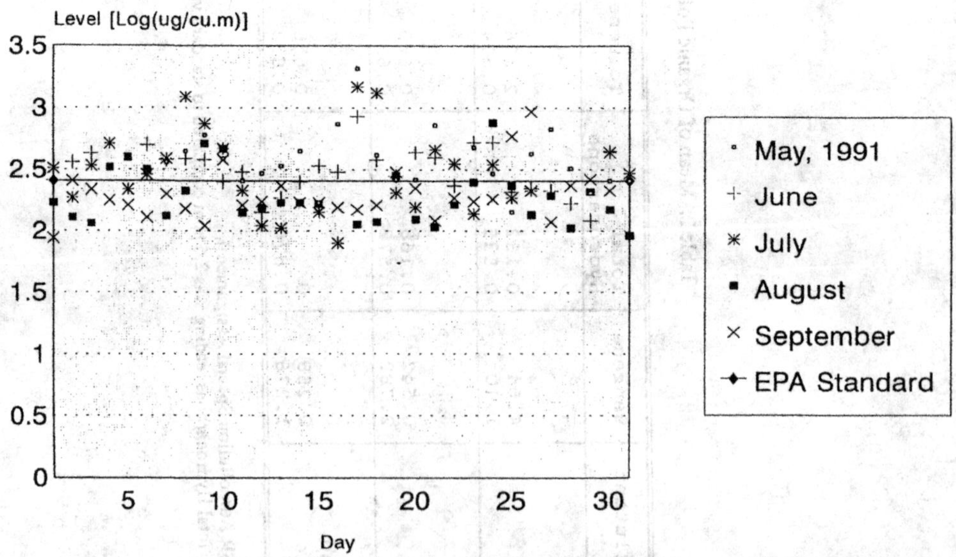

Modified From EPC Final Reports and Al-Majed et.al., 1991

Figure 4. Outdoor levels of suspended particulate matter (< 10 um) Al-Riqqa area

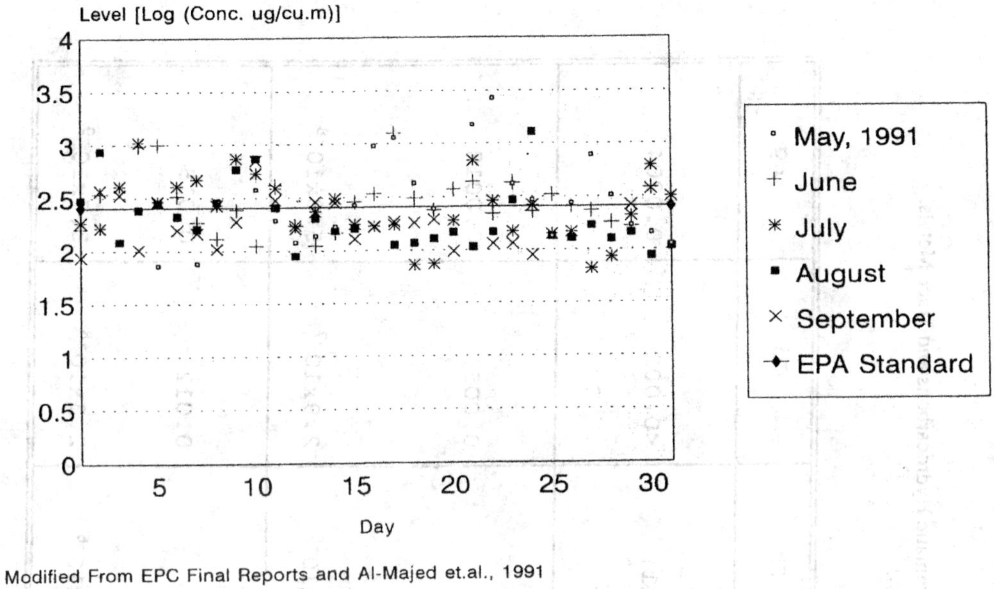

Figure 5. Outdoor levels of suspended particulate matter (< 10 um) Al-Jahra area.

obenzene, diethyl phthalate, dimethyl phthalate and naphthalene. These compounds can harm the liver, kidneys and respiratory system (The New York Times, International, 1991).

Abdali (1991) reported the results of analyses of indoor particulate matter by using charcoal filters from indoor-air purifier systems. Lead had the highest concentration of metals (0.62-1.1 mg/filter) and significant concentrations of Ni were also observed in these filters as shown in Table 3A and 3B. Nickel is significant due to its involvement in cancer. The organic composition of these particulates was also studied, benzene, naphthalene and aliphatic hydrocarbons were detected significantly as shown in Table 3B.

The joint study executed by KISR and DIER (Dubai Institute of Environmental Research) 1991, concluded that respirable particulates were high. High concentration of respirable dust (≤10 μm) was obtained using Cassella, U.K. a Portable Respirable Dust monitor. The fine particles of the dust ranged from 375 μg/m^3 (Kuwait City) to 585 μg/m^3 (Al-Ahmadi). Dust samples collected from air conditioner (AC) filters in Al-Ahmadi area showed about 50% different organic compounds when compared to oil sample and one possible source could be a product of environmental transformation of particulates during transport from the source to the exposure site. High concentration of metals such as Pb, Mn, Ni and Zn were detected in the AC filters as shown in Table 4. These elements are toxic and exposure to them is detrimental.

Besides the harm to the respiratory systems, particulate matter input to water tanks (reservoirs) poses a serious health concern. In the presence of free chlorine, trihalomethanes (THM's) may possibly form from the reaction with polyaromatic hydrocarbons adsorbed onto the particulate matter. Another potential threat is the aerial deposition of soot on the surface of vegetation and soil which will create a continuous source of pollution during the resuspension of particulate contaminants or consumption of local farm produce.

Table 3. Preliminary Data for the Analysis of the Air Purifier for Nuclear Aromatic Hydrocarbons and Trace Metals. (Sampled in June 1991 in Kuwait City.)

A. Trace Metals

Sample	V	Ni	Pb	Cd	Hg
Blank Sample					
mg/filter (A)	<0.004	<0.003	0.011	<0.002	0.0004
mg/m^3	–	–	–	–	–
48 hr. sample					
mg/filter (B)	0.030	0.048	0.62	0.005	0.0018
15th floor- Meridien Hotel					
mg/m^3	$1.7x10^{-6}$	$2.8x10^{-6}$	$3.5x10^{-5}$	$2.9x10^{-7}$	$5.8x10^{-8}$
2 months sample					
mg/filter (C)	0.084	0.11	1.1	0.012	0.015
2nd floor Al-Shuwaik					
mg/m^3	$1.6x10^{-7}$	$2.1x10^{-7}$	$2.1x10^{-6}$	$2.3x10^{-8}$	$5.7x10^{-9}$

B. Polyaromatic Hydrocarbons

Sample	C(10)	C(11)	C(12)	C(11)	C(12)	C(11)
Blank mg/filter mg/m^3	3 –	0.07 –	3 –	0.6 –	0.5 –	0.4 –
48 hours sample 15th floor Meridien Hotel mg/filter mg/m^3	139.72 8.0×10^{-3}	244.8 14×10^{-3}	448.8 26×10^{-3}	918 53×10^{-3}	1346 78×10^{-3}	143 8.2×10^{-3}
2 months sample 2nd floor Al-Shuwaik mg/filter mg/m^3	214 4×10^{-4}	257 5×10^{-4}	349 7×10^{-4}	612 1.2×10^{-3}	857 1.7×10^{-3}	92 1.8×10^{-4}

Sample	Cyclohexane + C(n)		Benzene + C(n)	
	n=4	n=5	n=3	n=4
Blank mg/filter mg/m^3	2 –	0.08 –	ND –	ND –
48 hours sample mg/filter mg/m^3	33 1.9×10^{-3}	13×10^{-3} 13×10^{-3}	2000 115×10^{-3}	510 29×10^{-3}
2 months sample mg/filter mg/m^3	27 5.2×10^{-5}	141 3×10^{-4}	979 2×10^{-3}	324 6.2×1^{-4}

Sample	Naphthalene	Methyl Naphthalene	Dimethyl Naphthalene
Blank mg/filter mg/m^3	ND –	2 –	1 –
48 hours sample mg/filter mg/m^3	8.16 5×10^{-4}	2.24 1.3×10^{-4}	1.24 7.1×10^{-5}
2 months sample mg/filter mg/m^3	4.65 9×10^{-6}	1.1 2.1×10^{-6}	1 1.9×10^{-6}

Table 4. Heavy Metals Concentration in μg/g in Dust Samples Collected from Air Conditioner Filters.

Sample No.	Cd	Cu	Co	Fe	Pb	Mn	Ni	Zn
AC-1	ND	103.73	ND	91.581	306.375	467.643	166.708	602.715
AC-2	ND	64.828	ND	57.293	250.692	341.221	181.455	519.097
AC-3	ND	186.309	ND	63.433	368.607	362.997	154.795	2194.60

Source: DIER/KISR, Joint Study 1991

Soil samples were analyzed throughout the joint activity of DIER and KISR. Several organic peaks were detected and further analyses are required for qualitative identification. Moreover, from the preliminary analyses more peaks were detected in the soil surface than in 6″ deep soil in Al-Ahmadi and Kuwait City and oil pool coastal sample, creating a potential and permanent sink for PAHs in the environment. Nevertheless, the further the deposition took place from the wells the lower the organic concentrations observed (contamination at Al-Ahmadi > Kuwait City). Metal concentrations varied in concentration, for example, Pb (5.0–9.0 μg/g); Mn (69–107 μg/g) and Ni (12–22 μg/g).

Benzo(a)pyrene, often used as an indicator of health impact of PAHs, has been detected both in the white smoke and in the ambient dust (Mulholland et al, 1991). The Japanese team also detected this compound in Kuwait City air samples and the concentration was higher than in Japan. (Japanese Report, 1991).

Johnson et al. (1991) reported differences in smoke particle morphology. The particles were composed of spherules of 0.1 μm diameter forming chains of hundreds in number and several microns in length (180) km at the end of March, 1991. However, samples from 1000 km from Kuwait had different appearance, the spherules were more likely to be closely packed into near spherical clusters, although they have the same sizes. Their measurement of the density of smoke was 1 g/cm^3. Graedel et al. (1991) reported an estimate of the initial plume. They observed that the soot concentration was 770 μg/m^3 in March and 640 μg/^{3} in July.

Al-Majed (1991) reported that fluoranthene and pyrene were among the highest concentrations in PAH species adsorbed on air particulate matter. The range of concentrations was (0.36–71.35 ng/m^3) and (6.11–40.11 ng/m^3) for fluoranthene and pyrene, respectively. Other carcinogenic PAHs were detected at lower concentrations such as benzo(a)pyrene (0.08–30.57 ng/m^3), benzo(a)anthracene (0.05–35.74 ng/m^3), benzo(a)fluoranthene (0.07–44.29 ng/m^3), and dibenzo (a,h) anthracene (0.04–50.90 ng/m^3). In the 3 samples analyzed at Al-Mansoriah, of the 10 PAHs detected, about 58.41% belong to the non-carcinogenic group, 7.4% belong to the carcinogenic group and 34.2% belong to the strongly carcinogenic group. Most of the maximum levels of the PAH's were higher than the maximum permissible limit of 1 ng/m^3 as included in the USSR Report, (1991). Furthermore, the Japanese and French standard levels, are also lower than the detected levels in Kuwait. The 1989 Japanese maximum level is 15.2 ng/m^3, and the average is 2.7 ng/m^3 (Japanese Report, 1991) and the French maximum level in winter 1988 is 17.6 ng/m^3 (French Report, 1991).

From the EPC report (1991) we observed that as the inhalable (<10 μm) particulate matter concentration decreases, benzo(a)pyrene concentration increases. This was indicated in the PM_{10} samples compared to Hi-volume samples. As shown in Figures 6a to c, the effects of minimum and maximum particulate matter from the PM_{10} samples on the organic concentration (such as extractable organic matter (EOM), chrysene equivalent, benzo(a)anthracene (B(a)A), benzo(a)pyrene (B(a)P), benzo(g,h,i)pyrelene (B(g,h,i)P), dibenzo-(a,h)anthracene (Db(a,h)A), and total polyaromatic hydrocarbons (TPAHs)) are different from the ones from the Hi-volume samples (Al-Majed, 1991). This adsorption behaviour of PAHs raises a public health concern, due to their association with particles of < 10 μm. This is due to the fact that as the excess presence of binding sites reduces the PAHs contact sites on the inhalable particulate matter, the capacity of B(a)P adsorption is increased on the available sites. This phenomenon is also observed in the study of Mulholland et al. (1991) study, where, in the plume sample, B(a)P concentration was 19 μg/g compared to 2.7 μg/g in the ambient sample (see Table 5). The particulate concentrations in the plume and ambient samples were 18,000 $\mu g/m^3$ and 600 $\mu g/m^3$ giving rations of (0.001 and 0.004), respectively.

Japanese results showed 1014 $\mu g/m^3$ total suspended particulate concentrations and particles of <10 μm size of 0.952 mg/m^3 or 952 $\mu g/m^3$. These concentrations which are higher than the EPA standard level.

Thousands of tons of toxic compounds were pollution our environment British Report (1991) and MEPA Report (1991) among them were the high concentration of suspended particulates (estimated at approx. 1500 ton/day) which could be a direct and indirect source of pollution for the marine ecosystem. Trace organic compounds, such as chlorinated and petroleum hydrocarbons, have been shown to be major and persistent marine pollutants in the near-shore marine ecosystem along the coastal plain of southern California, an oil-producing region, (Young et al., 1975; 1978; 1980; 1983; 1988 and 1991).

Based on the examination of the volume transport budget, it has been indicated that water residence time in the Arabian Gulf is about 2–5 years on average. Thus, water flushing and dispersal in the Gulf is relatively slow, as compared with more tidally and advectively energetic seas, such as the North Sea. Due to the high salinity of the Arabian Gulf, distribution and partitioning of oil into the marine system is affected. Abdali (1990) showed that the concentration of oil and its sulfur polyaromatics in seawater on a laboratory scale was increased by increasing the salinity from 10 to 50 $^0/_{00}$. Therefore, the amount of oil expected to enter the water in solution will be higher due to the salting-in effect by implementing the Setschenow equation [($KCs=\log_{10} S_0/S$) where, S_0 and S are the solubilities in moles/L in distilled water and in salt solution, the latter with molarity Cs].

It is expected that the spilled oil and the atmospheric fallout of pollutants from the smoke of the damaged oil fields will cause pollution of the marine environment. We need to know which pollutants the fish are exposed to in the marine environment, which of these pollutants tend to accumulate, what is the fate of these pollutants inside the organism and what are the biological impact accompanying these exposures.

This pollution can have short-term and long-term effects. Short-term effects (acute) could affect the marine organisms, including fishes and shrimps. Long-term exposure (chronic) to a polluted marine environment will result in bioaccumulation of pollutants in the living organisms. Several studies have demonstrated that organisms living in or on polluted sediments can bioaccumulate contaminants (NRC, 1985; Varanasi, et al., 1985; NRC, 1983; NRC, 1981; McElroy, et al., 1987).

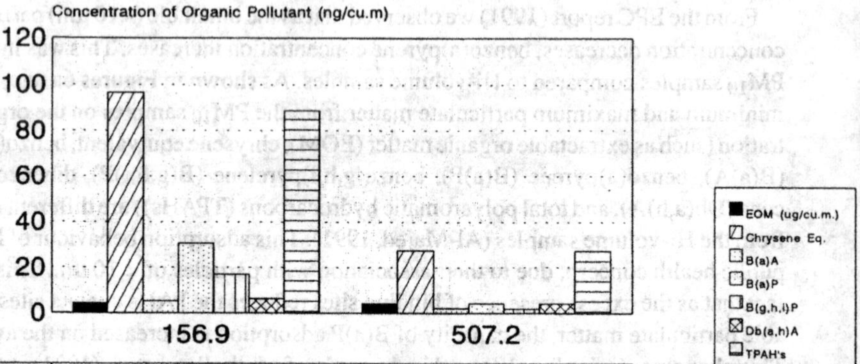

	156.9	507.2
EOM (ug/cu.m.)	4.28	4.16
Chrysene Eq.	96.48	27.52
B(a)A	5.18	2.02
B(a)P	30.7	4.19
B(g,h,i)P	17.09	1.83
Db(a,h)A	6.22	4.22
TPAH's	88.67	27.46

Minimum and Maximum Particulate Matter Concentration (ug/cu. m.)

Modified From EPC Final Reports and Al-Majed et.al., 1991

Figure 6a. Effect of maximum particulate concentration (PM–10) on PAH's adsorption, Al-Mansouriah area.

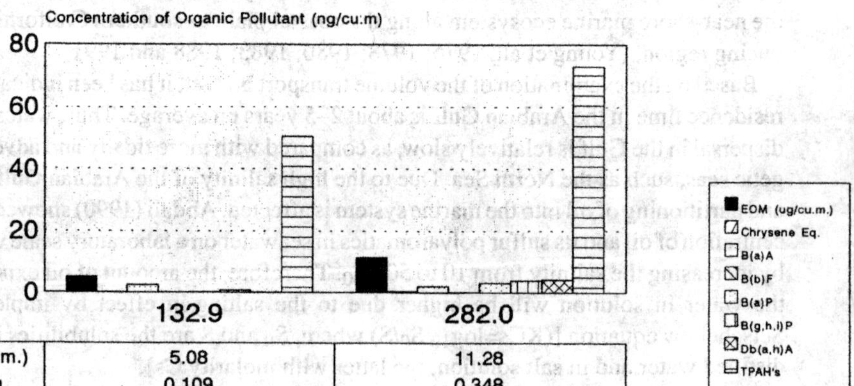

	132.9	282.0
EOM (ug/cu.m.)	5.08	11.28
Chrysene Eq.	0.109	0.348
B(a)A	2.43	2.07
B(b)F	0.09	0.1
B(a)P	0.17	3.07
B(g,h,i)P	0.66	3.85
Db(a,h)A	0.04	4.17
TPAH's	50.58	73.03

Minimum and Maximum Particulate Matter Concentration (ug/cu. m.)

Modified From EPC Final Reports and Al-Majed et.al., 1991

Figure 6b. Effect of maximum particulate concentration (PM–10) on PAH's adsorption, Mubarak Hospital.

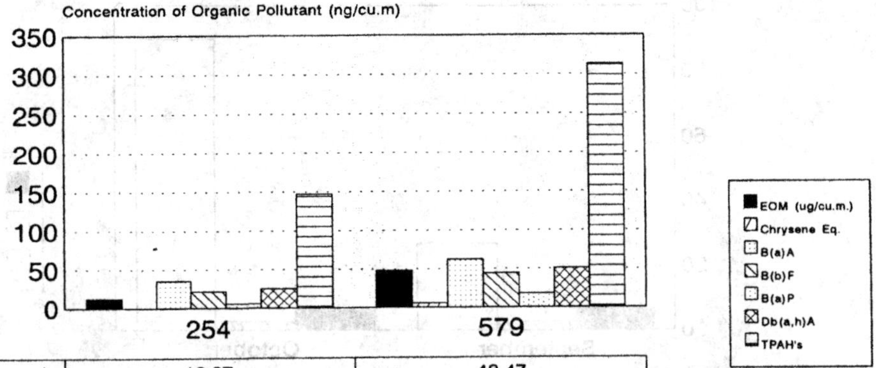

	254	579
EOM (ug/cu.m.)	12.97	48.47
Chrysene Eq.	0.203	5.36
B(a)A	35.74	62.14
B(b)F	21.76	44.29
B(a)P	5.79	17.66
Db(a,h)A	25.34	50.9
TPAH's	147.16	312.95

Minimum and Maximum Particulate Matter Concentration (ug/cu. m.)

Modified From EPC Final Reports and Al-Majed et.al., 1991

Figure 6c. Effect of maximum particulate concentration (PM–10) on PAH's adsorption, Shwaikh area.

Table 5. Physical and Chemical Properties of Plume, Ambient and Lab Scale Samples

Property	Plume	Ambient	Lab scale
Particulate Concentrations	15-21 mg/m^3	0.6±0.3 mg/m^3 (600±300 μg/m^3)	100-200 mg/m^3
CO_2 concentrations	650 ppm	350 ppm	1000-2000 ppm
smoke yield	0.11 g smoke/g oil	–	0.09-0.15 g sm./g or 1
organic fraction	92+3%	72±5%	14-21%
Total PAH/g sample	490 μg (16 μg/m^3) Kuwait whitish smoke)	50 μg (Ambient dust st.)	1400 μg
benzo(a) pyrene/g sample	19 μg (0.3 μg/m^3) (Kuwait whitish smoke)	2.7 μg (Ambient dust st.)	180-240 μg

Source: Mulholland et al., 1991

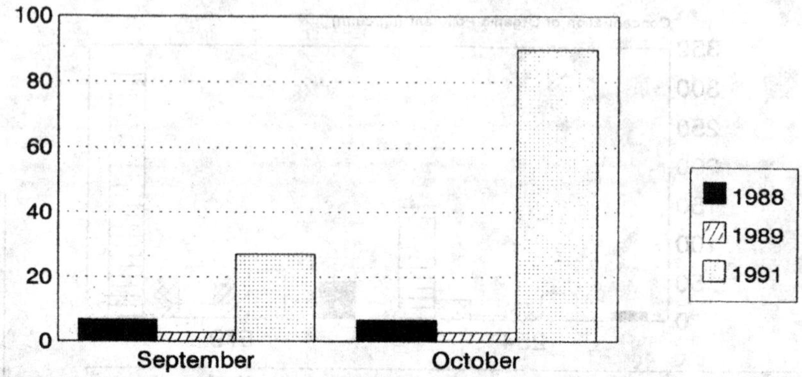

	September	October
1988	6.88	6.71
1989	2.69	2.78
1991	27.04	89.8

Source: EPC Final Report, 1991

Figure 7. Mean value of petroleum hydrocarbons (µg/L) in Kuwaiti coastal waters (September–October 1988, 1989, 1991).

The most hazardous pollutants related to the present environmental crisis in Kuwait are the polyaromatic compounds of crude oil as well as some heavy metals. Satellite imagery results suggest that depositional products falling in the Arabian Gulf are reaching levels of concern which will adversely impact the marine ecosystem even after the mitigation of effects of oil spill (TASC, 1991). Figure 7 summarises the mean value of petroleum hydrocarbons in Kuwaiti coastal waters during September-October, 1988–1991 (EPC, 1991). The concentration during 1991 was significantly higher than previous years. Whether a compound persists in the environment, degrades or bioaccumulates, is a function of the physical and chemical properties of the compound, and its interaction with the physical, chemical and biological elements of the ecosystem that influence its fate and toxicity. Therefore, PAHs pose a potential serious threat to health, because they are among the lipophilic compounds that have high resistance to biodegradation and are not readily excreted from the exposed contaminated organisms and hence, could transfer from lower levels of the food chain to top predators and subsequently to humans. Consequently, the KISR/EES Division initiated a study in July, 1991 to determine the levels of these pollutants in fishes and shrimp in Kuwait, (Abdali and Literathy, 1991).

It is noteworthy that short-term acute oil pollution, when the marine environment is exposed to high oil concentrations such as in the vicinity of significant oil spills, will taint the meat of fishes and shrimp and will alert the consumer about the possible contamination of that food because of its odor and the presence of oil. In addition, regular analysis of the fishes and shrimp can ensure quality characterization and advise people if the concentration of contaminants exceeds acceptable level.

Eight species of fish and one shrimp species were collected at local fish market and were analyzed for heavy metals and petroleum hydrocarbons in the VITUKI Laboratory in Budapest, according to a contractual agreement between KISR and VITUKI signed in July

1991. Table (6) lists the fish and shrimp species sampled for this study. Tables 7a to f summarise the concentrations of heavy metals and polyaromatic hydrocarbons in these samples. Almost all the detected levels of heavy metals are higher than the minimum world ranges, however, rarely exceeded the maximum world ranges.

Overall, there is a slight increase in the PAH concentrations, especially in SU, HA, NG, NW and SR samples. This leads us to a preliminary conclusion that the accumulation of the PAHs in the fat is correlated with the body size, since these sampled fishes were of medium to large size. The shrimp in sampling segment 3 (September) were larger than in the previous two segments of sampling (July and August).

The detected B(a)P was below the food quality standard value (500 ng/kg wet-weight) (Tables 7d-f). The concentrations were directly correlated with the amount of total petroleum hydrocarbons as shown in SR-1, SM-2, HA-1, NG-2 in Table 7d, ST-1, Sm-3, TH-1 and NW-1 in Table 7e and SU and NW in Table 7f.

A study conducted by Varanasi and Stein (1991) concluded that high molecular weight PAHs such as B(a)P, would rarely be detected because of extensive metabolism. Varanasi et al. (1989) suggested the determination of the distribution of these metabolites in the fish body by using radio-labelled hydrocarbons. Stein, et al. (1987) concluded that these metabolites are mainly concentrated in the hepatobiliary systems rather than edible muscles. Therefore, in the long-term impact studies, it is recommended to also analyze bile, liver and skin for PAHs.

Moreover, there are exceptions for this distribution. If the hepatic cytochrome P-450 levels in sexually mature fish was affected (for instance, when the organism is acutely exposed to an oil spill) and hence its ability to metabolize and excrete PAHs, was reduced the concentration of PAHs in the extrahepatic tissues and edible muscles may be reduced. (Buhler and Williams, 1989; Varanasi et al., 1982; Collier et al., 1986).

The Human Health Concern:

Many oil field fires were located north and south of Kuwait City (Figure 8). Approximately 752 wells were sabotaged. Of these 81% were on fire and 5.7% were gushing (Kuwait Oil Wells Blow-out, Aspect and Effects, 1991), and 6–8 million barrels/day were continuously lost at first. The behaviour of the plumes is complicated and mainly controlled by meteorology. It was estimated that about 5–15% of the burned oil was emitted as soot, 250 tons/day CO, 65,000 tons/day SO_2 and 3000 tons/day NO_x. From the assumption made by Graedel et al. (1991), about 10^{12} g/day of Kuwaiti oil was being burned (which is higher than the Kuwaiti wells normal operation: 2×10^{11} g/day). The sulphur components of oil comprise about 2.5% by weight and, therefore, the estimated emissions of sulphur was 2.5×10^{10} g/day.

The environmental damage affected almost all the environmental compartments in Kuwait and the surrounded areas, however the air was obviously and significantly polluted. Among all the environmental impacts that are likely to result from the massive release of heavy black smoke and gases are human health effects, especially because a human being inhales daily about 15 Kg of air as compared with 2 Kg of water and 700 gm of solid food. Results of comparisons of particulate sizes with the exposure levels of PAHs' showed that, the inhalable fraction (< 10μm) (PM_{10}) contains higher PAH's than the total particulate matter (>10μm) (Hi-volume), as shown in Figure 9 and the ratios of TPAH's/Chrysene equivalent (for organic matter) were higher in PM_{10} samples than in Hi-volume samples. The

Table 6. Fish and Shrimp Species Sampled.

Local Name	Species Name	Common Name	Code
Hamoor	Epithelus tauvina	Brown-spotted grouper	HA
Sobaity	Acanthopaqrus cuvien	Silvery black porgy (seabream)	ST
Sheem	Eleutheronema tetradactylum	Four-threads threadfin	SM
Nagroor	Pomadasys argenteus	Sivery grant	NG
Zobaidy	Pampus argenteus	Silvery pomfret	ZO
Suboor	Hilsa ilisha	River shad	SU
Newaiby	Otolithes argenteus	Silvery croaker	NW
Thelah	Clorinemus lysan	Spotted leather	TH
Shrimp (Raubyane Om-Nairah)	Penaues semisulcatus	Grooved tiger prawn	SR

Table 7a. Concentration of Selected Heavy Metals in Fish Samples from Local Fish Markets in Kuwait, Collected in July, 1991.

Sample		Heavy Metal Concentration, ug/g Wet Weight				
No	I.D.*	Hg	Cd	Pb	Ni	V
1	SU-1	0.17	0.011	0.22	0.34	0.08
2	SU-2	0.36	0.010	0.28	0.17	0.05
3	SU-3	0.15	0.015	0.18	0.25	0.04
4	ST-1	0.52	0.022	0.20	0.35	ND
5	ST-2	0.38	0.019	0.18	0.27	ND
6	ST-3	0.32	0.019	0.16	0.26	ND
7	SM-1	0.12	0.025	0.23	1.20	ND
8	SM-2	0.42	0.015	0.24	0.26	ND
9	SM-3	0.15	0.017	0.15	0.44	ND
10	HA-1	0.35	0.011	0.08	0.29	0.04
11	HA-2	0.27	0.010	0.09	0.39	ND
12	HA-3	0.16	0.008	0.39	0.22	ND
13	NG-1	0.15	0.024	0.44	0.23	ND
14	NG-2	0.23	0.020	0.20	0.26	ND
15	NG-3	0.23	0.023	0.23	0.34	ND
16	ZO-1	0.03	0.037	0.30	0.43	0.09
17	ZO-2	0.13	0.016	0.10	0.23	ND
18	ZO-3	0.05	0.012	0.21	0.22 -	ND
19	SR-1	0.03	0.260	0.26	0.28	0.03
20	SR-2	0.23	0.078	0.11	0.28	0.03
21	SR-3	0.18	0.094	0.22	0.35	ND
Detection Limit		0.01	0.001	0.02	0.02	0.02
World range		0.01-0.5	0.001-0.35	0.02-0.40	0.01-0.50	0.01-0.10

ND= Not Detectable

* SU= Suboor, ST= Sobaity, SM= Sheem, HA= Hamoor, NG= Nagroor, ZO= Zobaidy, SR= Shrimp

Table 7b. Concentration of Selected Heavy Metals in Fish Samples from Local Fish Markets in Kuwait, Collected in August, 1991.

Samples		Heavy Metal Concentration, μg/g Wet Weight				
No	I.D.*	Hg	Cd	Pb	Ni	V
1	SU-1	0.012	0.017	0.273	0.188	0.026
2	SU-2	0.018	0.014	0.224	0.211	0.040
3	SU-3	0.012	0.020	0.149	0.212	0.040
4	ST-1	0.206	0.015	0.188	0.137	0.009
5	ST-2	0.098	0.035	0.074	0.467	0.020
6	SM-1	0.937	0.021	0.041	0.313	ND
7	SM-2	1.580	0.014	0.285	0.256	0.014
8	SM-3	0.610	0.017	0.252	0.170	0.012
9	HA-1	0.242	0.054	0.720	0.245	ND
10	HA-2	0.167	0.020	0.242	0.273	ND
11	HA-3	0.233	0.023	0.211	0.136	ND
12	NG-1	0.124	0.021	0.191	0.267	0.012
13	NG-2	0.132	0.027	0.118	0.349	0.015
14	NG-3	0.151	0.029	0.169	0.443	ND
15	ZO-1	0.030	0.022	0.473	0.283	ND
16	ZO-2	0.038	0.029	0.197	0.258	ND
17	ZO-3	0.042	0.021	0.088	0.469	ND
18	TH-1	0.383	0.024	0.253	0.167	ND
19	TH-2	0.728	0.029	0.030	0.428	ND
20	TH-3	0.814	0.011	0.094	0.315	ND
21	NW-1	0.290	0.032	0.168	0.238	ND
Detection Limit		0.01	0.001	0.02	0.02	0.02
World range		0.01-0.5	0.001-0.35	0.02-0.4	0.01-0.5	0.01-0.10

ND= Not Detectable

* SU= Suboor, ST= Sobaity, SM= Sheem, HA= Hamoor, NG= Nagroor, ZO= Zobaidy, TH=Thela, NW= Newaiby.

Table 7c. Concentration of Selected Heavy Metals in Fish Samples from Local Fish Markets in Kuwait, Collected in September, 1991.

Sample No	I.D.*	Hg	Cd	Pb	Ni	V
		Heavy Metal Concentration, ug/g Wet Weight				
1	SU	0.020	0.010	0.064	0.177	0.062
2	ST	0.400	0.009	0.020	0.126	ND
3	SM	0.500	0.013	0.111	0.093	0.013
4	HA	0.415	0.010	0.045	0.153	0.009
5	NG	0.130	0.011	0.077	0.147	0.016
6	ZO	0.075	0.022	0.063	0.166	ND
7	NW	0.280	0.012	0.048	1.105	0.015
8	SR	0.030	0.087	0.203	0.148	0.015
Detection Limit		0.01	0.001	0.02	0.02	0.02
World range		0.01-0.5	0.001-0.35	0.02-0.40	0.01-0.50	0.01-0.10

ND= Not Detectable

* SU= Suboor, ST= Sobaity, SM= Sheem, HA= Hamoor, NG= Nagroor, ZO= Zobaidy, NW=Newaiby, SR= Shrimp

Table 7d. Concentration of Total Petroleum Hydrocarbons Determined with Fluorescence Spectroscopy (ROPME Method) as well as Benzo(a)pyrene (BaP) with GC/MS in Fish Samples from Local Fish Market in Kuwait, Collected in July, 1991.

Sample		Total Petroleum Hydrocarbons, μg/g wet weight expressed in ROPME Oil Standard		
		Ex/Em		BaP
No.	I.D.*	310/360	390/415	ng/Kg
1	SU-1	3.78	10.3	ND
2	SU-2	1.31	5.4	ND
3	SU-3	0.26	5.4	ND
4	ST-1	1.51	1.1	ND
5	ST-2	5.36	4.7	ND
6	ST-3	0.66	0.9	125
7	SM-1	3.35	4.5	ND
8	SM-2	2.05	4.4	3
9	SM-3	1.37	4.0	ND
10	HA-1	0.57	2.0	1
11	HA-2	0.97	1.3	ND
12	HA-3	0.60	0.9	ND
13	NG-1	0.22	0.9	ND
14	NG-2	0.16	3.4	3
15	NG-3	0.59	2.3	ND
16	ZO-1	0.97	2.1	19
17	ZO-2	1.48	2.0	ND
18	ZO-3	1.35	1.6	ND
19	SR-1	0.44	1.2	6
20	SR-2	0.19	0.2	ND
21	SR-3	0.07	0.2	ND

ND= Not Detectable
* SU= Suboor, ST= Sobaity, SM= Sheem, HA= Hamoor, NG= Nagroor, ZO= Zobaidy, SR= Shrimp.

Table 7e. Concentration of Total Petroleum Hydrocarbons Determined with Fluorescence Spectroscopy (ROPME Method) as well as Benzo(a)pyrene (BaP) with GC/MS in Fish Samples from Local Fish Market in Kuwait, Collected in August, 1991.

Samples		Total Petroleum Hydrocarbons, µg/g wet weight expressed in ROPME Oil Standard		
		EX/EM		Bap
No	I.D.*	310/360	390/415	ng/kg
1	SU-1	1.80	4.8	ND
2	SU-2	2.60	4.0	ND
3	SU-3	2.00	6.3	ND
4	ST-1	2.15	5.0	5
5	ST-2	1.80	3.2	ND
6	SM-1	3.20	5.2	ND
7	SM-2	1.10	4.8	ND
8	SM-3	2.50	6.0	3
9	HA-1	0.50	1.8	ND
10	HA-2	1.30	2.0	ND
11	HA-3	0.46	1.5	ND
12	NG-1	0.21	0.8	ND
13	NG-2	0.25	1.2	ND
14	NG-3	0.22	1.2	ND
15	ZO-1	0.69	2.8	ND
16	ZO-2	1.15	2.5	ND
17	ZO-3	1.05	1.6	ND
18	TH-1	2.40	5.5	3
19	TH-2	1.80	4.8	ND
20	TH-3	1.90	4.0	ND
21	NW-1	4.00	10.2	5

ND = Not Detectable

* SU= Suboor, ST= Sobaity, SM= Sheem, HA= Hamoor, NG= Nagroor, ZO= Zobaidy, TH= Thela, NW= Newaiby.

Table 7f. Concentration of Total Petroleum Hydrocarbons Determined with Fluorescence Spectroscopy (ROPME Method) as well as Benzo(a)pyrene (BaP) with GC/MS in Fish Samples from Local Fish Market in Kuwait, Collected in September, 1991.

Samples		Total Petroleum Hydrocarbons, µg/g wet weight expressed in ROPME Oil Standard		
		EX/EM		Bap
No	I.D.*	310/360	390/415	ng/kg
1	SU	4.20	8.60	15
2	ST	1.50	2.20	ND
3	SM	1.00	1.30	ND
4	HA	1.20	2.20	ND
5	NG	2.50	5.40	ND
6	ZO	0.90	0.70	ND
7	NW	5.00	6.50	5
8	SR	0.80	0.80	ND

ND = Not Detectable

* SU= Suboor, ST= Sobaity, SM= Sheem, HA= Hamoor, NG= Nagroor, ZO= Zobaidy, NW= Newaiby, SR= Shrimp.

particulate composition is more important than the particulate weight. Therefore, using PM_{10} results are more reliable for exposure and risk assessment.

Compared to other pollution disasters, (Table 8) it is needless to say that the meteorological and geographical factors were extremely favourable in the environmental disaster in Kuwait. These two factors played a very important role in the dispersion of pollutants and the dilution of the concentrations at the exposure sites and finally the mitigation of the impact on the ecosystems including human being.

Physicians in Kuwait feel that they observed an increase in the cases of asthma and non-reversible chronic obstructive lung disease (WMO, 1991). They also observed an irritation of eyes and upper air passages. Local residents were surveyed during April, 1991 and they have complained of headaches and eye irritation (Abdali and Khuraibet, 1991). Table 9 summarizes the increase in some diseases after the oil fires. The Analytic Sciences Corporation (TASC, 1991) estimated a 10% increase in the sickness symptoms due to the fires after two months of exposure. Total patients from "Al-Adan Hospital" showed an increase in respira-

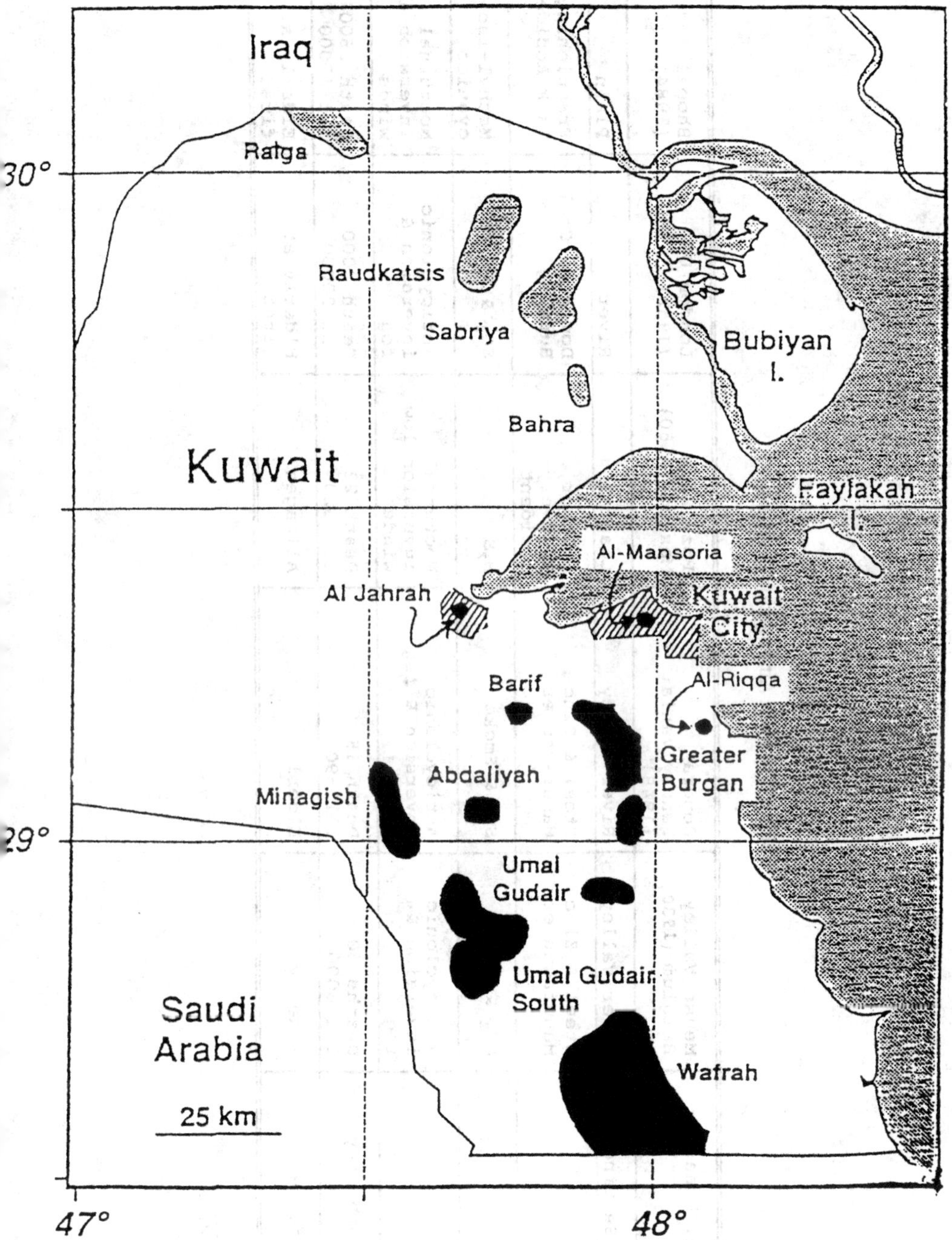

Figure 8. Kuwait main oil fields.

Table 8. The Main Air Pollution Disasters

Air Poll. Disaster / Parameter	Meuse Valley Belgium (1930)	Donora Pennsylvania (1948)	Poza Rica Mexico (1950)	London UK (1952)	Bhopal India (1984)
Geographical Setting	River Valley	River Valley	Coastal	River	Plain
Source	Steel & Zinc Manufacturer	Steel & Zinc Manufacturer	Sulphur Recovery-accident	Domestic Coal Burning	Fractioning of Tank Accident
Pollutants	SO_2 & Smoke	SO_2 & Smoke	H_2S	SO_2 & Smoke	Methyl-iso-cyanide
Weather	Anticyclonic inversion & fog	Anticyclonic inversion & fog	Nocturnal inversion low winds	Anticyclonic inversion & fog	Nocturnal inversion low winds
Mortality & Morbidity	Deaths 60 ill 6000	Death 15 ill 5900	Death 22 ill 320	Death 4000 ill 20000	Death 2500 ill 10000
Age Groups Affected	Elderly	Elderly	All ages	Elderly at first	Elderly at first

Table 9. Emergency Room Surveillance—Kuwait City

Conditions	Before Oil Fires	After Oil Fires
Hypertension	4.1%	8.3%
Heart diseases	2.8%	5.2%
Dizziness	0.08%	1.1%
Drug allergy	0.04%	0.5%
Chronic bronchitis	0.04%	0.5%
Gastrointestinal disorders	19.6%	23.2%

US Public Health Service Report, June 1991

tory incidence for 15.9–28.3% and in asthma for 5.7–6.7% from 1986 and 1991 (EPC, 1991).

The joint study between KISR and DIER concluded that the incidence of obstructive restrictive or mixed type of lung dysfunction was statistically higher (24 out of 34 cases or –70%) and were more common in females than in males. However, because of the small number of samples, more systematic studies on health impact are required before any final conclusion can be made.

Monthly statistics have been collected in Al-Ahmadi Hospital during the period April-September 1991 for respiratory problems. As shown in Table 10 the incidence rate of respiratory illness (% of total number of patients) after exposure to pollution is significantly higher than cases before the exposure (1987, 1988 and 1989) and this may result from exposure to smoke and gases emitted from burning oil wells.

Although the initial pollution disaster has been, short-lived and within those 9 months the number of burning oil wells was diminishing, the exposure was unlike the known long-term (chronic) exposure trends in pollution in industrialized countries. However, some of the long-term effect of toxic chemicals may occur as delayed effects of short-term exposure to high doses, (Tyler, 1990).

A complete exposure analysis program is required to define the total exposure and dose from all media (soil, food, water, seafood, air and plants) through all the routes of exposure (inhalation, ingestion and dermal). This goal requires a multidisciplinary expertise of environmental scientists, toxicologists, molecular biologist, engineers, modellers, statisticians, epidemiologists, physicians, industrial hygienists and social scientists. Exposure analysis is an extremely important field and is critical today in order to provide guidelines for its application in risk assessment and risk management. The prevention of environmentally related diseases by establishment of exposure-dose-effect relationships is necessary to define the exposure reduction strategies for those populations at risk.

BIO-MONITORING FOR ENVIRONMENTAL HEALTH IMPACT ASSESSMENT

One way to detect the long-term impact is to use biomonitoring methodologies which could be indicated by DNA lesions resulting from the interaction of some pollutants (such as the

Table 10. Incidence Rate of Respiratory Illness Before and After Exposure to Pollution 1991

Months	1987	1988	1989	Ave. 3 years	1991
April	3.67	3.46	3.15	3.43	4.16
May	3.60	3.05	2.60	3.08	4.07
June	2.44	2.28	2.69	2.47	3.73
July	2.42	2.45	2.65	2.51	4.25
August	2.08	2.49	2.53	2.37	3.64
September	2.33	2.76	2.81	2.63	4.68
Total	16.49	16.49	16.43	16.49	24.53
P<0.01					

Source: Dr. M. W. Daoud, 1991.

products of the incomplete combustion of oil fires) with the genetic material of individuals exposed. Biomonitoring is a modern technology and due to ethical considerations in performing research on humans and the limitation in analytical methodologies, it has been the least implemented element in a control program. The lesions may act as initiators of somatic disease such as cancer and inherited disease such as congenital abnormalities. Three steps are required for this task:

1. Qualitative and quantitative analysis of pollutants at the exposure sites.
2. Identification of exposed individuals.
3. Quantitative analysis of DNA and genetic damage in exposed tissues.

Once the biological effects of the exposure are determined, the data base from the above tasks will be used to predict the potential somatic and genetic risks of the participants involved in the study and the general population of Kuwait. Moreover, this will provide a good foundation for long term epidemiological studies of the population of Kuwait and probably of the Gulf.

This type of laboratory examination of the instances of chemical intoxication is helpful not only in the treatment of an affected individual, but in the prevention of further similar occurrences as well as in the establishment of liability for a chemical exposure.

Choosing biological specimens is a very critical task and is affected by several variables such as the metabolic profile of the specific chemical, availability of published reference limits, route of exposure, time of sampling, availability of the specimen and analytical method to be implemented.

One example of the biomonitoring indicator is the concentration of benzene in urine and a blood specimen could be used for routine monitoring purposes. Following human exposure to benzene, only about 12% of a dose is exhaled unchanged from the lungs and about 0.1% is excreted unchanged in the urine. The remainder is metabolized in the liver to highly toxic oxidation products, such as phenolic metabolites (phenol, catechol and hydroquinone) (Baselt, 1988).

Monitoring lead in the urine is another example of a biomonitoring indicator. Over 90% of absorbed lead is deposited in bone, primarily in dense bone, with only minor amounts excreted in hair, nails or urine. Many diagnostic tests are based on the abnormalities that result from increases in erythrocyte zinc protoporphyrin, urinary coproporphyrin and urinary-aminolevulinic acid (ALA) and inhibition of erythrocyte ALA dehydrase, which are indication for lead poisoning. ALA in urine is related to deleterious effects of absorbed lead and the standard level for ALA is 4.5 mg/g creatinine. However, the blood lead level is another index for lead toxicity, where the standard level should not exceed 0.7 to 0.8 mg/L (Baselt, 1988).

Nickel, one of the toxic metals, is believed to be normally excreted in urine with an average concentration of 4.5 µg/L (Andersen, et al., 1978). Insoluble nickel compounds that are inhaled tend to accumulate in the nasal mucosa and lungs and nickel levels in these tissues may remain elevated for years after cessation of exposure (Torijussen and Andersen, 1979). Urine and blood specimens are good indicators for nickel contamination.

There are insufficient data concerning the applicability of using biomonitoring indicator for PAH-induced effect. Therefore, it is recommended that a modified methodology for PAHs, be developed as it seems to be the main human health concern based upon the preliminary data discussed in the environmental monitoring data base.

References

Abdali, F. (1990). Studies on the Photolytic Behaviour of Dibenzothiophene in Crude Oil Water Systems. Doctor of Philosophy dissertation (Environmental Health Sciences), University of Michigan.

Abdali, F. (1991). Indoor Pollution Air Filters Analysis. Technical Report.

Abdali, F. and Literathy P. (1992). Burning Oil Wells in Kuwait: Their Potential Impact on the Environment and Human Health. Phase (1): Fish Quality Task Force. Final Report /ENV–3, Jan.

Abdali, F. and Khuraibet A. (1991). Kuwait Oil Fires: A Public Health Concern. (Under Preparation).

Al-Majed, N. (1991). Assessment of Polycyclic Aromatic Hydrocarbons Levels in Air Suspended Particulate in the State of Kuwait. EPD. November.

Al-Yakoob, S. Abdali, F. and Abdal, Y. (1992). Assessment of Risk Associated with Air Pollutants Resulting from Kuwait Oil Fires. (Part 1: Hazard Identification). Final Report.

Andersen, I., Torjussen W. and Zachariansen H. (1978). Analysis for Nickel in Plasma and Urine by electrothermal Atomic Absorption Spectrometry, with Sample Preparation by Protein Precipitation. Clin. chem. 24: 1198–1202.

Baselt, R. C. (ed.) (1988). Biological Monitoring Methods for Industrial Chemical. Second Edition. PSG Publishing Company.

British Report (1991). Burning Oil Wells in Kuwait: Their Potential Impact on the Environment and Human Health. May.

Buhler, D. R. and Williams, D. E. (1989). Enzymes Involved in Metabolism of PAH by Fishes and Other Aquatic Animals: Oxidative Enzymes (or Phase I Enzymes). In: *Metabolism of Polycyclic Aromatic Hydrocarbons in the Aquatic Environment*, (U. Varanaci, Ed.), CRC Press, Boca Raton, FL, pp. 151–184.

Collier, T. K., Stein, J. E., Wallace, R. I., and Varanasi, U. (1986). Xenobiotic Metabolizing Enzymes in Spawning English Sole (Parophrys vetulus) Exposed to Organic Solvent Extracts of Marine Sediments from Contaminated and Reference Areas. Comp. Biochem. Physiol. 84C: pp. 291–298.

Daoud M. W. (1991). Ahmadi City Under the Smoke.

Dubai Institute of Environmental Research (DIER) A Report of Joint Study (1991), Dubai UAE and Kuwait Institute for Scientific Research (KISR) Kuwait. Preliminary Survey of Environmental Contamination and its Impact in Kuwait Following Fire in Oil Fields.

Environmental Protection Council (EPC) (1991). Final Reports, June-September.

EPA, Air Quality Guidelines. In: *Environmental Quality*. The Eleventh Annual Report of the Council on Environmental Quality.

French Report (1991). AIRPARIF Final Report. Measurement Campaign of the Regional Mobile Laboratory for Measurement of Air Quality in Kuwait. AIRPARIF, Laboratoire Central de la Prefecuture de Police, Atomic Superior Hygien Council, Laboratory Hygien of Paris City, and Analyses of French Institute of Petroleum. May.

Graedel, T. E., Baer, N. S., Francy J. P., Penner, J. E. and Sinclair J. D. (1991). AT&T Report. Degradation of Materials and Cultural Property by Emittants from Burning Kuwaiti Oil. Technical Memorandum, May.

Gulf Regional Air Monitoring Program (1991). Saudi Arabia, Action Plan. April.

Harvard Conference Abstracts (1991). The Kuwait Oil Fires. Conclusions and Recommendations. Sponsored by: The Arab Fund for Economics and Social Development, and the United nations Development Program. MA/USA. 12–14 August.

Japanese Report (August, 1991). Air Pollution from the Kuwait Oil Fires. Japanese Official Mission on the Environmental and Health Effects of the Oil Fires.

Johnson, D. W., Kilsby, C. G., McKenna, D. S., Saunders, R. W., Jenkins, G. J., Smith, F. B. and Foot, J. S. Airborne Observations of the Physical and Chemical Characteristics of the Kuwait Oil Smoke Plume. Meteorological Office. Meteorological Research Flight.

KISR Air Pollution Special Assignment. January, (1992). Technical Report (Under Preparation).

Kuwait Oil Wells Blow-out. Aspect and Effects (1991). KOC Report. MEPA Report. Environmental Impact of Oil Burning in the Kuwaiti Oil Fields. A Preliminary Evaluation. The research Institute/King Fahd University of Petroleum and Minerals, Dharan, Saudi Arabia.

Literathy, P. (1991). Personal Communication.

McElroy, A. E., Farnington, J. W. and Teal, J. M. (1987). Bioavailability of Polynuclear Aromatic Hydrocarbons in the Aquatic Environment. In: *Metabolism of Polynuclear Aromatic Hydrocarbons (PAHs) in the Environment*. (U. Varanasi, Ed.), CRC Press, Boca Raton, FL, pp. 1–39.

Mulholland, G. W., Benner, B. A., Fletcher, R. A., Steel, E., Wise, S. A., May, W. E., Madrzykowski, D., and Fwoin, D. (June, 1991). Analysis of Smoke Samples from Oil Well Fires in Kuwait. U.S. Department of Commerce, National Institute of Standards and Technology. Submitted to NOAA.

New York Times (1991). International, Tuesday, July 16.

NRC (1985). Oil in the Sea. Inputs, Fates and Effects, National Research Council, National Academy Press, Washington, D.C.

NRC (1983). Polycyclic Aromatic Hydrocarbons: Evaluation of Sources and Effects. National Research Council, National Academy Press, Washington, D.C.

NRC (1981). Testing for Effects of Chemicals on Ecosystems. National Research Council, National Academy Press, Washington, D.C.

Özkaynak, H. (1991). Measurements of Air Pollution from the Fires and Their Implications to Health. Presented at Darwin Scientific Foundation. Kuwait Oil Fires and International Environment Considerations, Washington Conference. September.

Stein, J. E., Hom, T., Casillas, E., Friedman, A., and Varanasi, U. (1987). Simultaneous Exposure of English Sole (Parophrys retulus) to Sediment Associated Xenobiotics. II. Chronic Exposure to an Urban Estuarine Sediment with Add ^{3}H benzo(a)pyrene and ^{14}C-Polychlorinated biphenyls . Mar. Environ. Res. 22:123–149.

TASC Report 1991. Janota, P., Chase R. R. P., Koplik, C. M. and Medler, C. L. (1991). Computer Model Assesses the Environmental Impact of Kuwait Oil Fires.

Torjussen, W. and Andersen I. (1979). Nickel Concentrations in Nasal Mucosa, Plasma, and Urine and Active and Retired Nickel Workers. Ann. Clin. Lab. Sci. 9: 289–298.

Tyler, G., Meller, Jr. (1990). Wad Sworth Publishing Co. Belmont, California pp. 450–482.

USSR Report 1991. Satunov and Nazarov. Proceedings of WMO meeting on the Effects of Kuwait Oil Fires.

Varanasi, U., Nishimoto, M., Reichert, W. L. and Stein, J. E. (1982). Metabolism and Subsequent Covalent Binding of Benzo(a)pyrene to Macromolecule in Gonads and Liver of Ripe English Sole (Parophyrs vetulus). Xenobiotica 12: pp. 417–425.

Varanasi, U., Stein, J. E., and Nishimoto, M. (1989). Biotransformation and Disposition of Polycyclic Aromatic Hydrocarbons (PAHs) in Fish. In: *Metabolism of Polycyclic Aromatic Hydrocarbon in the Aquatic Environment*. (U. Varanasi, Ed.), CRC Press, Boca Raton, Fl., pp. 93–149.

Varanasi, U., Reichert, W. L., Stein, J. E., Braun, D. W., and Sanborn, H. R. (1985). Bioactivity and Biotransformation of aromatic hydrocarbons. Environ, Sci. Technol. 19: 836–841.

WHO (1987). Air Quality Guidelines for Europe. WHO Regional Publications, European Series No. 23. World Health Organization Regional Office for Europe, Copenhagen.

WMO Report (1991). On the Meeting of Experts on the Atmospheric Part of the Joint U.N. Response to the Kuwait Oil Field Fires, Geneva. Executive Summary, 27–30 April.

WMO (1991) Response Report on Kuwait Oil Fires. April.

Young, D. R., McDermott, D. J. Heesen, T. C. and Jan, T. K (1975). Pollutant Inputs and Distributions off Southern California. In: *Marine Chemistry in the Coastal Environment* (T.M. Church, ed.). American Society Washington D.C., pp. 424–439.

Young, D. R., Jan, T. K. and Heesen, T. C. (1978). Cycling of Trace Metal and Chlorinated Hydrocarbon Wastes in the Southern California Bight. In *Estuarine Interactions* (M.L. Wiley, ed.). Academic Press, New York, pp. 481–496.

Young, D. R., Heesen, T. C. and Gossett, R. W. (1980). Chlorinated Benzenes in Southern California Municipal Wastewaters and submarine Discharge Zones. In *Water Chlorination-Environmental Impact and Health Effects, Vol. 3* (R. L. Jolley, W. A. Brungs, R. B. Cumming, and V. A. Jacobs, eds.). Ann Arbor Science Publ., Ann Arbor, MI, pp. 471–486.

Young, D. R., Gossett, R. W., Baird, R. B., Brown, D. A., Taylor, P. A. and Mille, M. J. (1983). Wastewater Inputs and Marine Bioaccumulation of Priority Pollutant Organics of Southern California. In *Water Chlorination-environmental Impact and Health Effects, Vol. 4* (R. L. Jolley, ed.). Ann Arbor Science Publ., Ann Arbor, MI, pp. 871–884.

Young, D. R., Gosett, R. W. and Heesen, T. C. (1988). Persistence of Chlorinated Hydrocarbon Contamination in a California Marine Ecosystem. In *Oceanic Processes in Marine Pollution, Vol. 5* (D. A. Wolfe and T. P. O'Connor, eds.) Krieger Publ., Malabar, FL, pp. 33–42.

Young, D. R., Mearns, A. J., and Gossett, R. W. (1991). Bioacccumulation of p,p'- DDE and PCB 1254 by a flatfish bioindicator from highly contaminated marine sediments of southern California. In *Organic Substances and Sediments in Water, Vol. 3* (R. Baker, ed.). Lewis Publ., Chelsea, MI, pp. 159–169.

THE IMPACT OF THE GULF WAR ON THE DESERT ECOSYSTEM*

Dr. FOZIA ALSDIRAWI

Aridland Agriculture Department
Food Resources Division
Kuwait Institute for Scientific Research
P.O. Box 24885, 13109–Safat–Kuwait

1. INTRODUCTION

The environmental crisis inflicted on Kuwait and this part of the world, by the Iraqi regime, during the assault on Kuwait from August 2, 1990 to the liberation day on February 26, 1991, has ramifications beyond human speculation or prediction. The demented exploding and burning of Kuwait's oil wells, the vehicular destruction of the land and the natural habitats, the placement of mines throughout, the coverage of vast areas with oil lakes and ponds, and the massive spillage of oil into the sea, have created environmental uncertainty of staggering proportions. In the absence of meaningful field scientific data, these impacts are collectively almost beyond human comprehension.

The aim of this presentation is to illustrate the different kind of threats to the desert ecosystem as created by the Iraqi army during their assault invasion of Kuwait. The presentation will also show the expected impacts of such threats to renewable natural resources of the desert ecosystem in Kuwait.

2. PRE-WAR STATUS OF THE DESERT ECOSYSTEM IN KUWAIT

Kuwait is politically, historically, and geographically part of Arabia, Arabian Gulf and the Western Palaearctic region, respectively (Harrison, 1964, 1968, 1977 and 1981; Bundy and Warr, 1980; Jennings, 1981; and Cramp et al., 1977, 1980, 1983, 1985 and 1988). Kuwait is an arid country with a harsh environment (Figure 1) and a very interesting and diverse biological heritage (Table 1).

Increasing both the human population (Figure 2) and the adaptation of technology for national growth (Figure 3) have contributed to alterations in Kuwait's natural ecosystems. The natural vegetation has been suffering tremendously from overgrazing, off-road ve-

*Workshop on the environmental Dimensions of the Gulf: Policy and Institutional Perspectives. Sponsored by U.A.E. University and the World Bank.

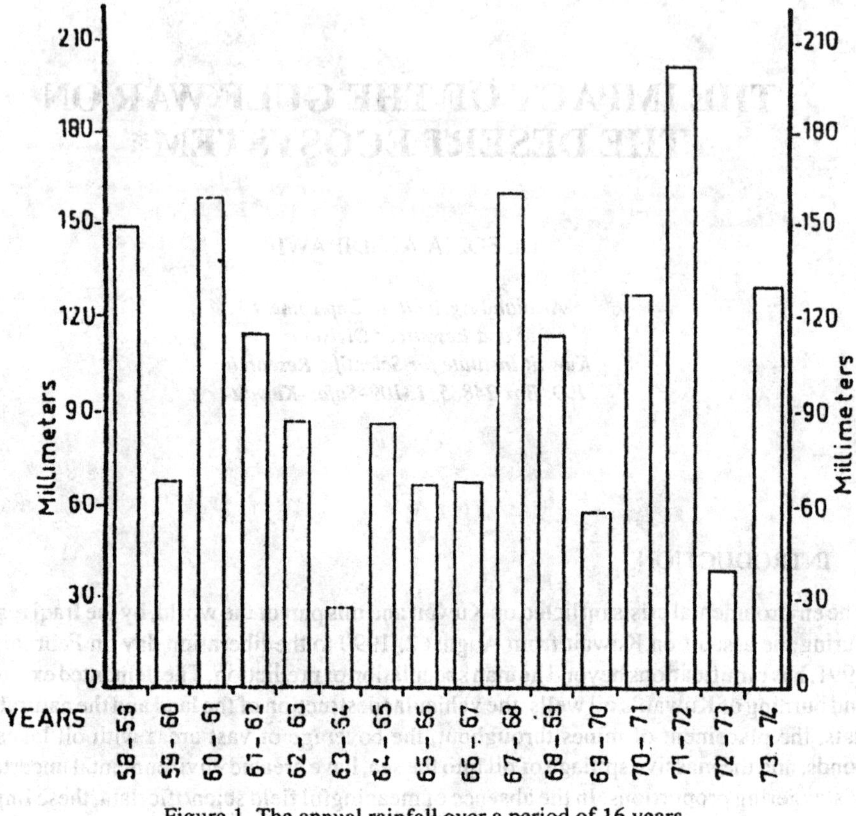

Figure 1. The annual rainfall over a period of 16 years.

hicle use and uprooting of woody shrubs for fuel consumptions (Alsdirawi, 1989). All these factors have contributed to the endangerment and disappearance of many wildlife and wild plants.

3. POST-WAR STATUS OF THE DESERT ECOSYSTEM IN KUWAIT

The Iraqi atrocity and aggression on the environment have left many scares, some of which might have an irreversible impact (Figure 4, 5, 6). The desert ecosystem was very productive (Figure 7, 8, and 9) compared to its physical and biological limitations.

The deliberate sabotaging of more than 700 oil wells have started Kuwait's inferno. The black smoke produced by the dirty burns of these wells covered the sun over an area of 15–30,000 km^2 (Figure 10). The clouds were constantly shifting with the wind direction. The total area that might experience soot fallout could cover hundreds of thousands of square kilometers of southern Iraq, western Iran, north-eastern Saudi Arabia as well as Kuwait. The ceiling height of the smoke cloud was about 5,000 meters, which might not of enough height to affect the stratospheric circulation, but it caused local convection effects and caused unseasonal winds and rain. The analyses of the smoke showed a concentration

Table 1. Biodiversity Characterizing Kuwait's Desert Biotope (Alsdirawi, 1989).

Mammals	**28 species**	
Extinct		8
Endangered		4
Rarely found		4
Unknown		6
Common		6
Birds	**300 species**	
Extinct		1
Former Breeders (FB)		50
Winter Visitors (WV)		47
Passage Migrants (PM)		130
WV/PM		62
WV/SV		1
Vagrant (V)		11
Reptiles	**40 species**	
Unknown		34
Rarely found		4
Vagrant (V)		2

of about 1,000 ppb. of sulphur dioxide and 50 ppb. nitrogen oxide at 2,000 altitude and 100 km from smoke. Sulphur levels were higher in the smoke from the southern Kuwait oil fields than from the northern oil fields (WCMC, 1991).

Particulate matter was recorded at 30,000 per c.c. over Kuwait. Considerable quantities of soot were deposited throughout the region with piles formed on the beaches and around the north and south oil fields. On the sea surface, streams of soot have produced an oily sheen on the surface of the water (WCMC, 1991). The plume threatened the crop and vegetation in Kuwait and neighbouring countries.

The oil spilled from the non-burning and gushing wells did not percolate into the sand, but drained into the creeks and wadis, forming oil lakes and ponds and causing more ecological damage (Figure 11). The sludge killed the upper life-forms by its toxicity and killed the deep life-forms by suffocation. The oil lakes were mistaken for water bodies by migra-

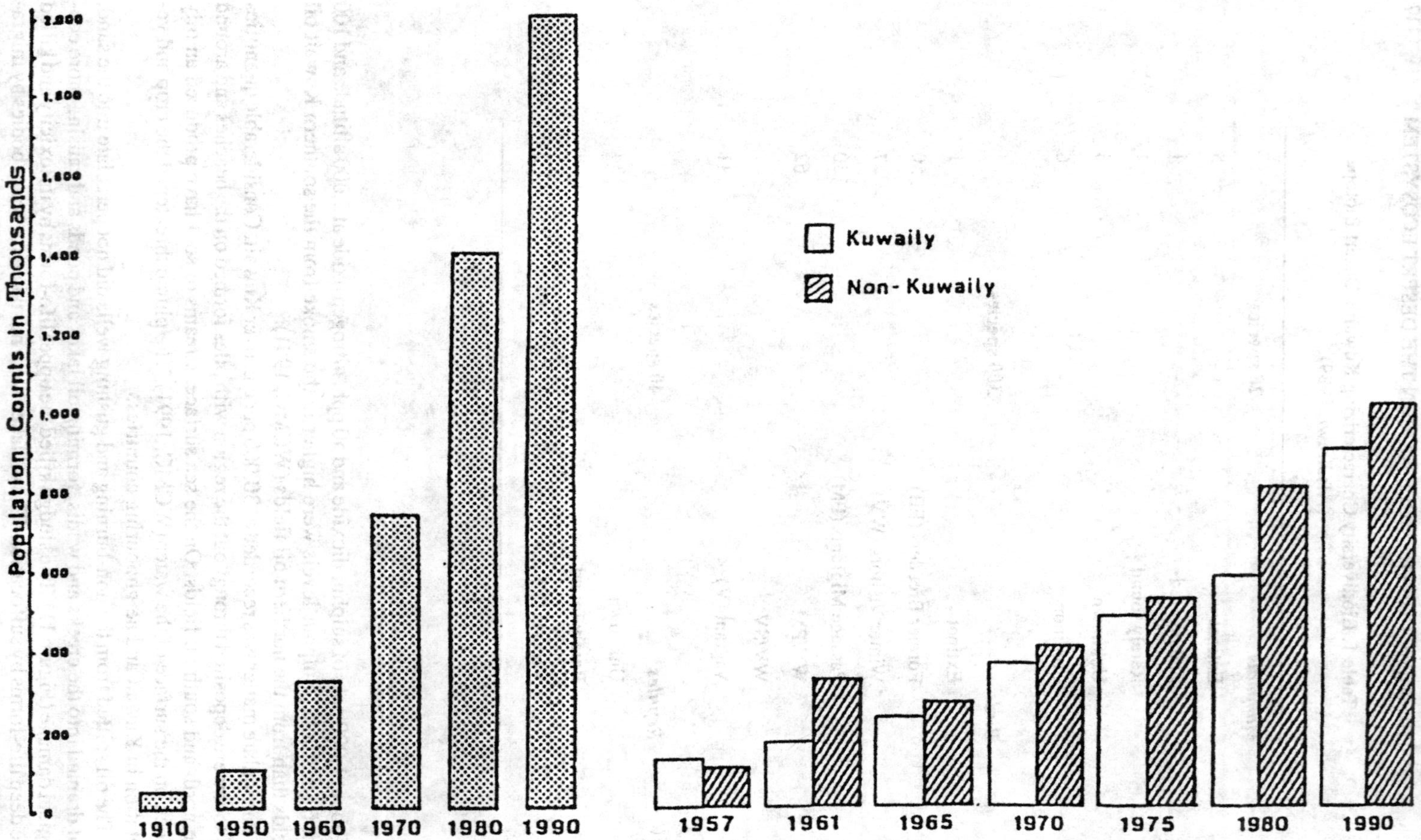

Figure 2. Population growth and demography of Kuwait from 1910 to 1990.

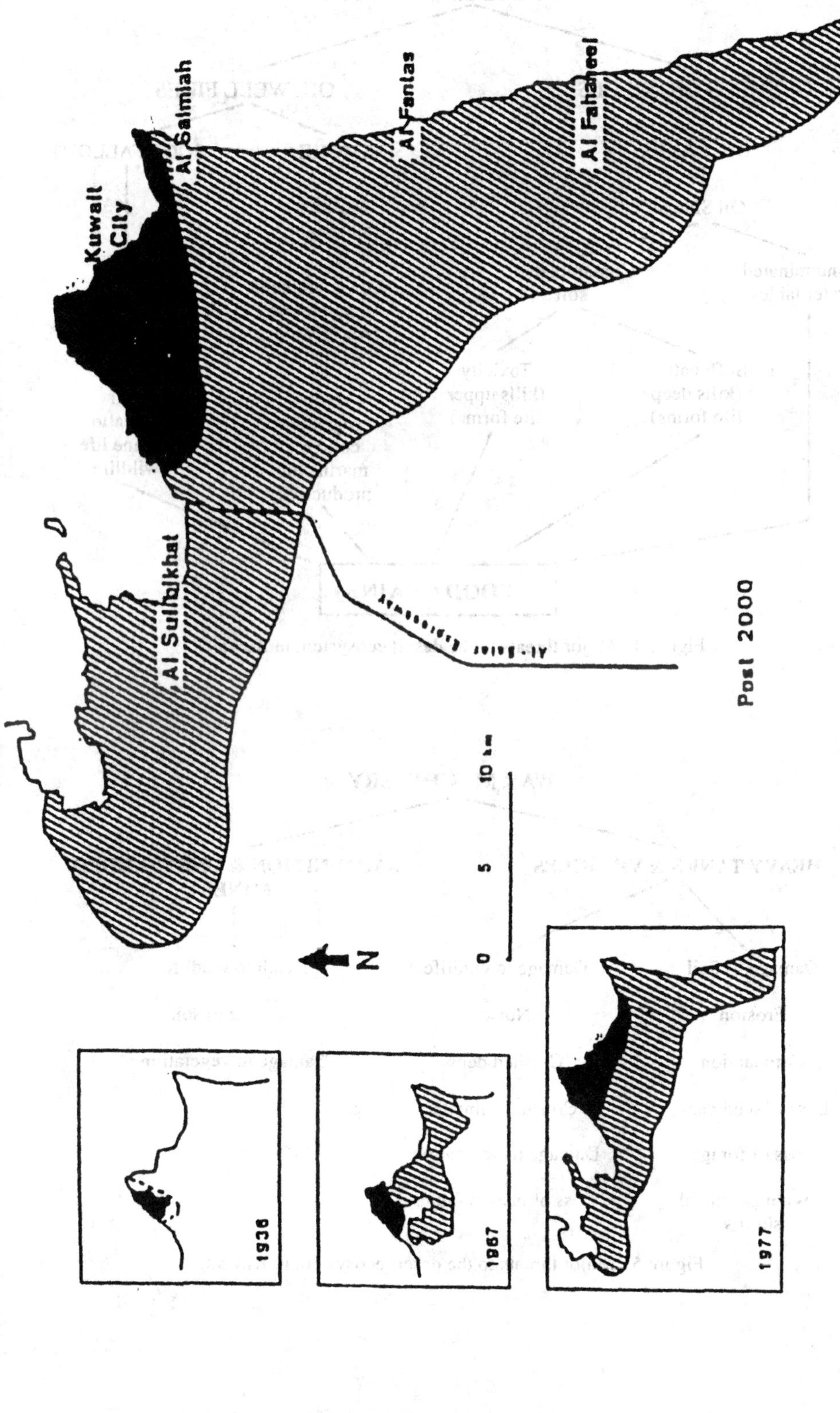

Figure 3. Urbanization growth map in Kuwait: past, current, and future.

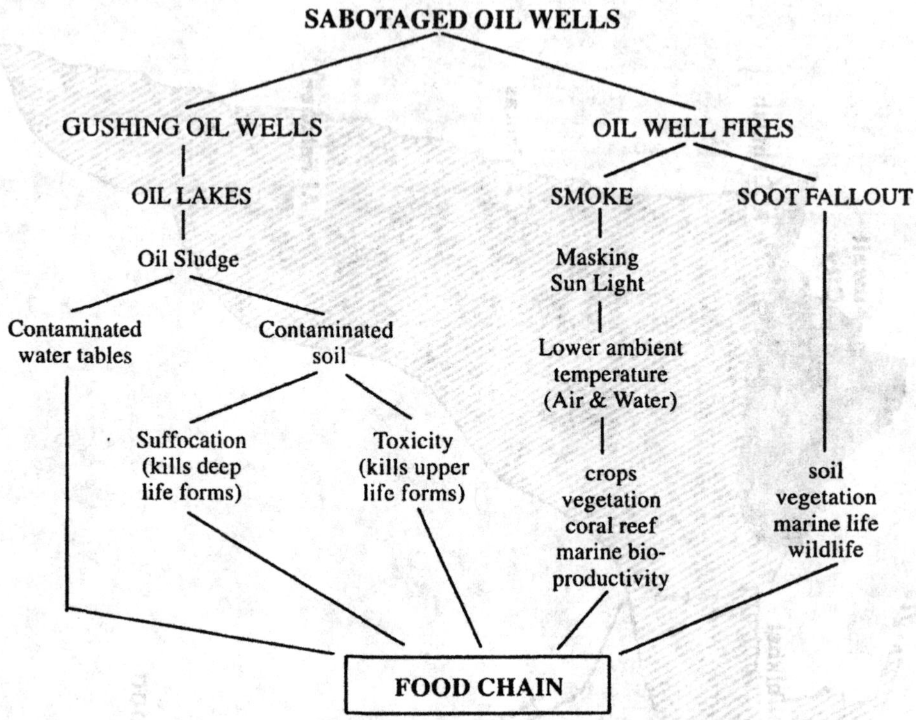

Figure 4. Major threats to the desert ecosystem in Kuwait.

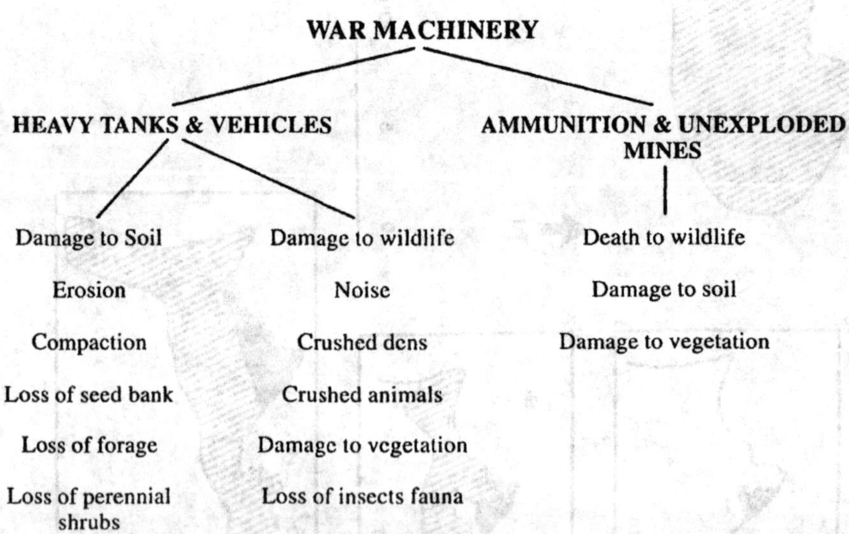

Figure 5. Major threats to the desert ecosystem in Kuwait.

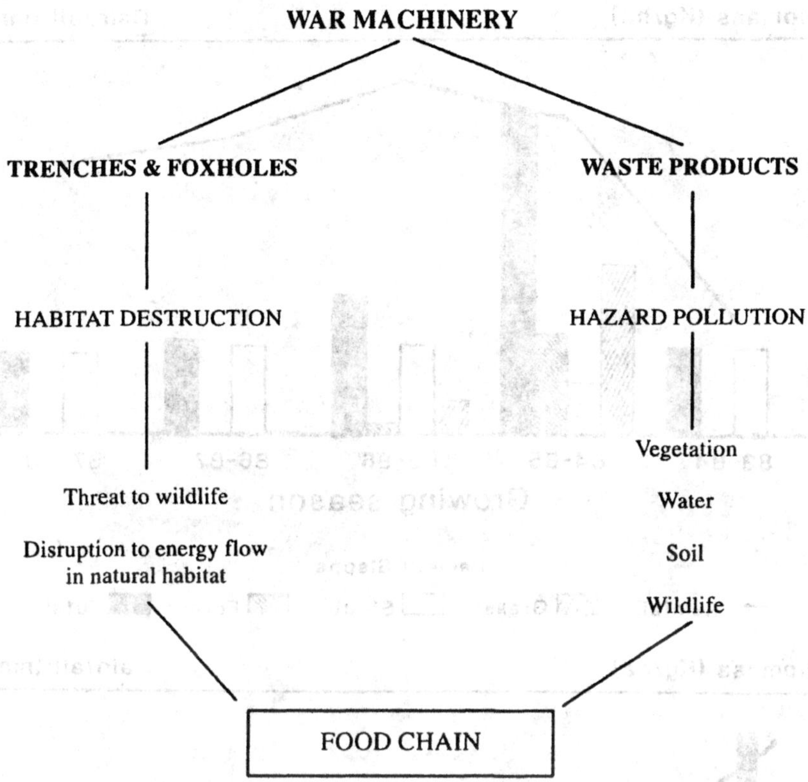

Figure 6. Threats to the desert ecosystem in Kuwait.

tory bird species as well as insects, both ended-up tragically with corposus scattered in and around the black death. The estimated figure of the crude oil in these lakes is around 60 million barrels (Bisharah, 1991 personal communication).

Tank tracks and deep ruts from tracked vehicles may take decades to disappear (Figure 12). Webb et al., 1983 reported that recovery of desert soils and vegetation from land use such as ORV activity will require at least decades and probably centuries. The direct impact of this threat is the breakage of the gravel and composition of the sand surface in the desert. This will increase the frequency of desert storms. Compression of the desert soil will preclude vegetation growth, especially during the germination season (February—early May). The noise of the war machinery scared the animals and flushed them from their natural habitats. A similar case was reported by (Brattstrom and Bondello, 1983) in California desert.

The unexploded mines and other ordnance in one hand, might deter graziers and hunters giving the habitats and wildlife an opportunity to recover. In the other hand, it will inflict a long-term threat to wildlife and will cause destruction of the vegetation and the soil.

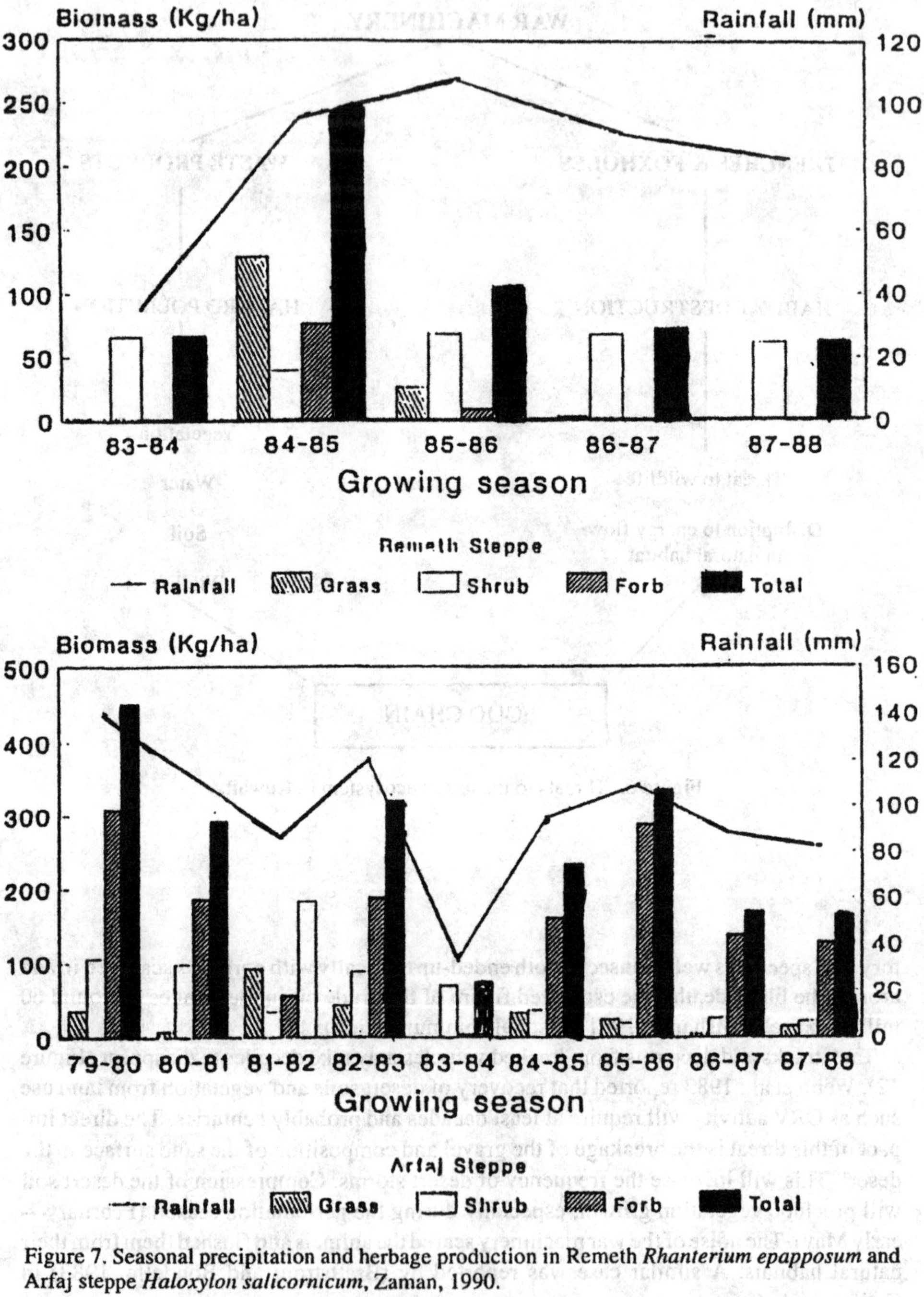

Figure 7. Seasonal precipitation and herbage production in Remeth *Rhanterium epapposum* and Arfaj steppe *Haloxylon salicornicum*, Zaman, 1990.

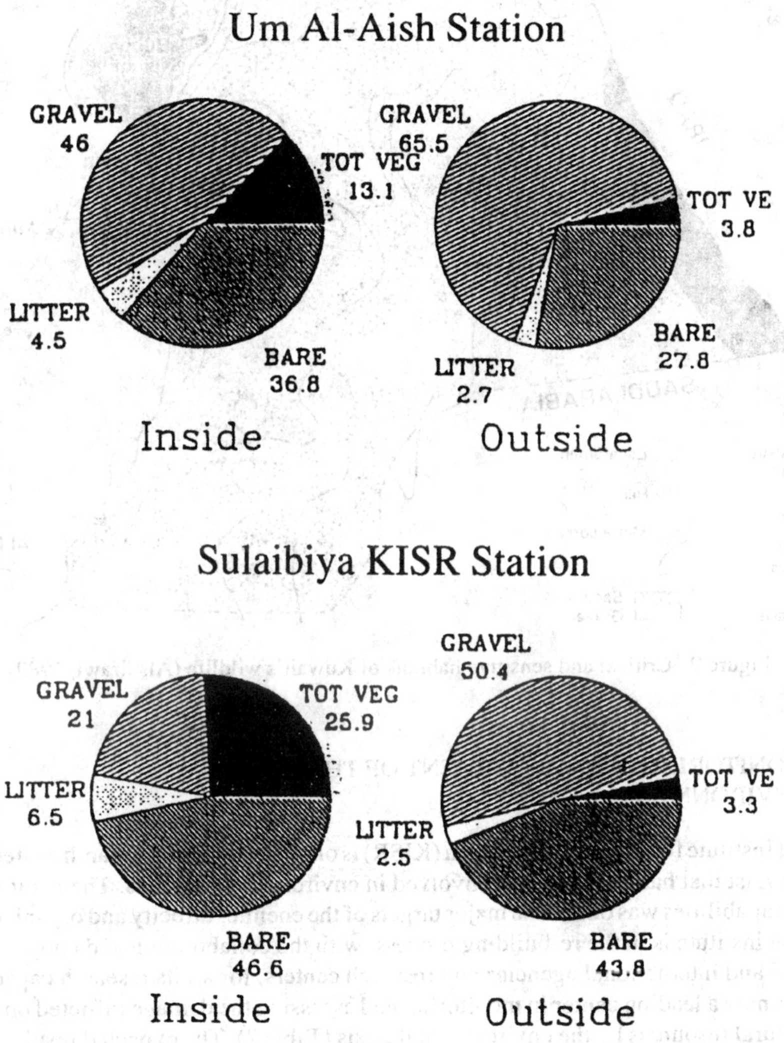

Figure 8. Average percent cover of the five growing seasons (1983–1988) inside and outside Umm Al-Aish station (Remeth steppe) and Sulaibiya KISR station (Arfaj steppe), Zaman, 1990.

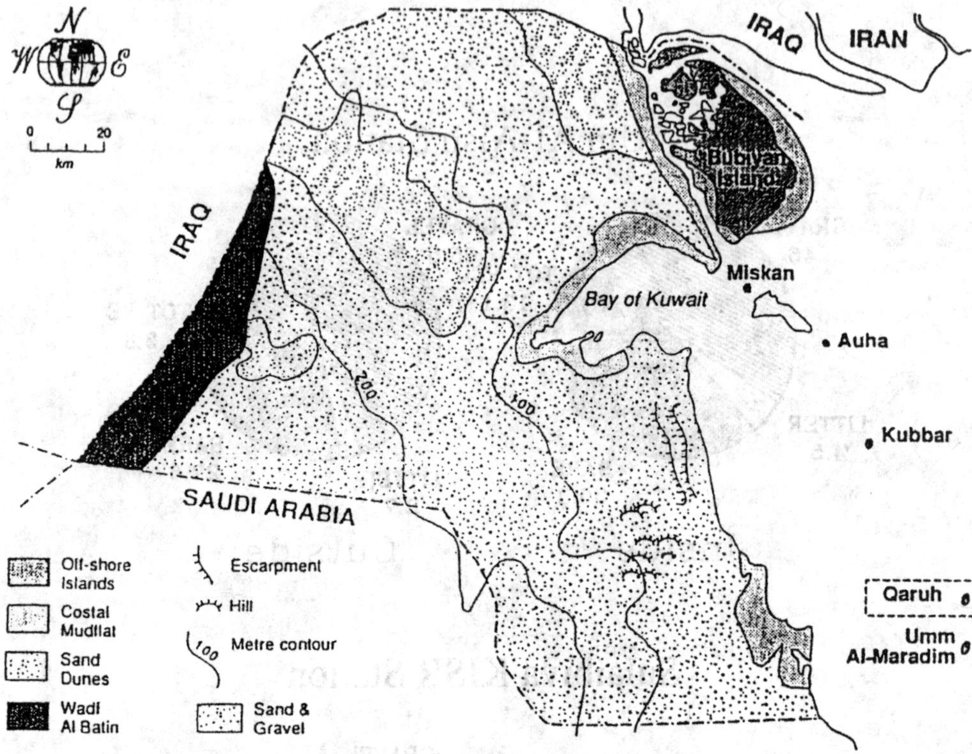

Figure 9. Critical and sensitive habitats of Kuwait's wildlife (Alsdirawi, 1989)

4. MONITORING AND ASSESSMENT OF THE ENVIRONMENTAL CRISIS

Kuwait Institute for Scientific Research (KISR) is one of the leading research centers in the Middle East that has been actively involved in environmental studies. The institute's research capabilities was one of the major targets of the enemies atrocity and organized looting. The institute is in the re-building process, with the collaboration and cooperation of regional and international agencies and research centers, for all its research capabilities. KISR is now a leading center in monitoring and assessing the damage inflicted on renewable natural resources by the environmental crisis (Table 2). The expected results of such activities will be published and presented in an international conference which will be hosted by KISR early 1993 in Kuwait.

International agencies with the cooperation of other government agencies, such as the Environmental Protection Council (EPC), Regional Organization for Protection of Marine Environment (ROPME) have been involved in the assessment of the environmental crisis. The Council of Ministers have allocated a land within the south oil-fields (Figure 13) to be used as a long-term monitoring environmental research station. The site will be the on-situ research center for all interested local, regional and international research bodies, Kuwait Oil Company (KOC) is the coordinator for the project.

Figure 10. A photograph showing oil wells' fire and smoke masking the sun. (COLOR PLATE V)

Figure 11. A photograph showing an oil lake.

Figure 12. A photograph showing the tracks of the war machinery in the open desert. (COLOR PLATE VI)

Table 2. List of short and long-term research conducted by Kuwait Institute for Scientific Research (KISR) to monitor and assess the impact of the environmental crisis on Kuwait's renewable natural resources.

1. Assessment of the impact of the crisis on groundwater pollution from massive oil spillage from damaged oil wells.
2. Assessment of the impact of oil pollution on feed water for, and product water from, desalination plants in Kuwait.
3. Assessment of the immediate damage to the fishery as a result of the Iraqi aggression.
4. Assessment of the effect of the oil spill and oil fire fallout on the plankton resources.
5. Long-term impact of the gulf crisis on the shrimp and fin fisheries.
6. Assessment of possible impact of oil pollution on the oyster beds of Kuwait.
7. Soil pollution by crude oil spill: The fate of PAH's in a field soil.
8. Effect of dry oil combustion pollutants on growth of certain native and indicator plants.
9. Assessment of the effect of oil spill and oil fire fallout on the desert renewable natural resources, wildlife and vegetation.
10. Analysis of farm animal tissues for detecting the impact of oil pollution on animal's health and suitability for human consumption.
11. Assessment of the pathological processes in feral cats exposed to Kuwait oil fires.
12. Rapid environmental impact assessment for the effects of the burning oil wells on Kuwait's total environment.
13. The distribution of emitted air pollutants from oil burning in the Kuwait oil fields.
14. Burning oil wells in Kuwait: Their potential impact on the environmental and human health, Phase I.
15. Environmental damage assessment of war activities to coastline of Kuwait.
16. Assessment of risk associated with air pollutants resulting from Kuwait oil fires.
17. Preliminary survey of environmental contamination and its impact in Kuwait following fire in oil fields.
18. Evaluation of drinking water quality of Kuwait.

Figure 13. A photograph showing an oil lake inside Jiadan environmental research center in the south oil fields, Kuwait.

5. MANAGEMENT RECOMMENDATIONS

- To establish a series of protected areas/National Parks as a systematic representation to all types of habitats and biodiversity characterizing the Gulf desert ecosystem, these systems must be supported by the legal and administration governmental supports.
- To establish a series of seed banks, captive breeding programs for the native wildlife and plants of the Gulf desert ecosystem.
- A restoration and remediation programs for the worsly hitted habitats within the desert ecosystem of the Gulf, using only environmentally sound approaches.
- To establish a regional awareness program addressing the environmental crisis of the Gulf, and to establish: international networks of information centers to advertise the crisis, the approaches used to address the problems, and the lessons learnt out of this crisis.
- To establish an *Environmental Crisis Rescues Center* in the Gulf to address all kinds of environmental crisis which might pass over the Gulf areas. Be sure to equip this center with highly qualified personnel and the right equipment and be sure that they are on the ulter all the time.
- To learn how nature deals with its wounds, let set aside pieces of land within the N. & S. oilfields of Kuwait as monitoring sites to watch the natural way of rehabitation of a highly impacted areas.

References

Alsdirawi, F., 1989. *Wildlife Resources of Kuwait: Historic Trends and Conservation Potentials*. Ph.D. Disseration, University of Arizona, USA.

Bisharah, J., 1991. Personal communication.

Brattstrom, B.H. and M.C. Bondello. 1983. Effects of off-road vehicle noise on desert vertebrates. **In:** R.H. Webb and H.G. Willshire (Eds), *Environmental Effects of Off-Road Vehicles. Impacts and management in arid regions*, New York, Springer-Verlag, pp. 167–206.

Bundy, G. and E. Warr, 1980. A check-list of the Birds of the Arabian Gulf States. *Sandgrouse* 1:4–49.

Cramp, S., K.E.L. Simmons, I.J. Ferguson-Lee, R. Gillmor, P.A.D. Hollom, R. Hudson, E.M. Nicholson, M.A. Ogilvie, P.J.S. Olney, K.H. Voos, and J. Wattel (eds.), 1977. *Handbook of the Birds of Europe, the Middle East and North Africa: The Birds of the Western Palaearctic*. Vol. I. Ostrich to ducks. Oxford University. Press, Oxford, England.

Cramp, S., K.E.L. Simmons, R. Gillmor, P.A.D. Hollom, R. Hudson, E.M. Nicholson, M.A. Ogilvie, P.J.S. Olney, C.B. Reselaar, K.N. Voous, D.I.M. Wallace, and J. Wattel (eds.), 1980. *Handbook of the Birds of Europe, the Middle East and North Africa: The Birds of the Western Palaearctic*. Vol. II. Hawks to bustards. Oxford Univ. Press, Oxford, England.

Cramp, S., K.E.L. Simmons, D.J. Brooks, N.J. Collar, E. Dunn, R. Gillmore, P.A.D. Hollom, R. Hudson, E.M. Nicholson, M.A. Ogilvie, P.J.S. Olney, C.S. Roselaar, K.H. Voous, D.I.M. Wallace, J. Wattel, and M.G. Wilson (eds.), 1983. *Handbook of the Birds of Europe, the Middle East and North Africa: The Birds of the Western Palaearctic*. Vol. II. Waders to gulls. Oxford Univ. Press, Oxford, England.

Cramp, S., D.J. Brooks, E. Dunn, R. Gillmor, P.A.D. Hollom, R. Hudson, E.M. Nicholson, M.A. Ogilvie, P.J.S. Olney, C.R. Roselaar, K.E. L. Simmons, K.H. Voous, D.I.M. Wallace, J. Wattel, and M.G. Wilson (eds.), 1985. *Handbook of the Birds of Europe, the Middle East and North Africa. The Birds of the Western Palaearctic*. Vol IV. Terns to woodpeckers, Oxford Univ. Press, Oxford, England.

Cramp, S., D.J. Brooks, E. Dunn, R. Gillmor, J. Hall-Graggs, P.A.D. Hollom, E.M. Nicholson, M.A. Ogilvie, C.R. Roselaar, P.J. Sellar, K.E.L. Simmons, K.H. Voous, D.I.M. Wallace, and M.G. Wilson (eds.), 1988. *Handbook of the Birds of Europe, the Middle East and North Africa. The Birds of the Western Palaearctic*. Vol. V. Flycatchers to thrushes, Oxford Univ. Press, Oxford, England.

Harrison, D., 1964. *The Mammals of Arabia*. Vol 1. E. Benn Ltd., London, England.

Harrison, D., 1968. *The Mammals of Arabia*. Vol 2. E. Benn Ltd., London, England.

Harrison, D., 1977. *The Mammals of Arabia*. Vol 3. E. Benn Ltd., London, England.

Harrison, D., 1981. *Mammals of the Arabian Gulf*. George Allen and Unwin, London, England.

Jennings, M.S., 1981. *Birds of the Arabian Gulf*. George Allen and Unwin, London, England.

WCMC., 1991. Gulf War environmental service series. World Conservation Monitoring Center. Cambridge, UK.

Webb, R.H., H.G. Wilshire, and M. Henry. 1983. Natural Recovery of soils and vegetation following human disturbance. **In:** R.H. Webb and H.G. Willshire (Eds), *Environmental Effects of Off-Road Vehicles. Impacts and management in arid regions*, New York, Springer-Verlag, pp. 279–302.

Zaman, S., 1990. Quantitative evaluation of vegetation in two main steppes of Kuwait's rangeland. A paper presented at the Second International Conference on Range Management in the Arabian Gulf, Kuwait, March 3–6, 1990. Kuwait Institute for Scientific Research, Kuwait.

GULF WAR DISRUPTION OF THE DESERT SURFACE IN KUWAIT

FAROUK EL-BAZ

Center for Remote Sensing
Boston University, Boston MA 02215

The Gulf War resulted in severe disruption of Kuwait's natural desert surface, which is here proposed as an ancient delta of the "Arabia River." Major disturbances were caused by the heavy traffic of military vehicles, the digging of trenches for soldiers, guns and ammunition, and the building of berms and other sand walls to hide military installations and/or hinder attacks. In affected areas, the result was the near total destruction of the desert pavement, a one-grain-thick layer of gravel that protects the fine-grained soil below from the action of wind. Mobilization of fine-grained soil increases the source area for dust storm activities in the short-term. It also increases the potential of sand dune formation and motion, with long-lasting detrimental effects on roads, farms and installations in the open desert; some dunes have already formed and began to encroach on roads since the cessation of military activities. Mitigation of the problems necessitates leveling the desert to its original flat contours, and stabilizing its fine soil through the rebuilding of a desert pavement and/or rejuvenating the growth of desert brush. Inaction or neglect will result in long-term environmental consequences of soil erosion, dust storm activities and the accumulation of sand in enough places and quantities to disrupt transportation and habitation sites in Kuwait and northeastern Saudi Arabia.

I. GENERAL SETTING

1. Major Geologic Features

Kuwait lies along the northwestern corner of the Arabian (Persian) Gulf, bordering Saudi Arabia to the south, Iraq to the north and west, and the Gulf to the east (Figure 1). The surface topography of its 17,818 square kilometers is marked by flat to undulating gravel-covered plains that slope gently towards the Gulf from a maximum elevation of 284 meters above mean sea level in the west.

The climate is arid to semi-arid with rainfall ranging during the past 30 years from less than 25 mm in 1964 to 375 mm in 1972, with an average of 115 mm per year. The precipitation infiltrates rapidly, leaving no permanent surface water, although some wadis fill temporarily with water from the winter rains. The largest of these is Wadi Al-Batin, which is 5 kilometers in width at the western border of the country. The author proposes that this wadi is part of a partially-buried, water-dug channel whose tributaries drain part of the Hijaz Mountains, 850 kilometers to the southwest in western Saudi Arabia. He further proposes to name the channel the "Arabia River"; its delta deposits form much of the land surface of Kuwait (Figure 2).

Geologically, Kuwait consists of flat-lying Tertiary rocks overlying gently folded Cretaceous and Jurassic formations. Exposed rock types include the Eocene Dammam Formation, a white, fine-grained cherty limestone that shows some karst development at its

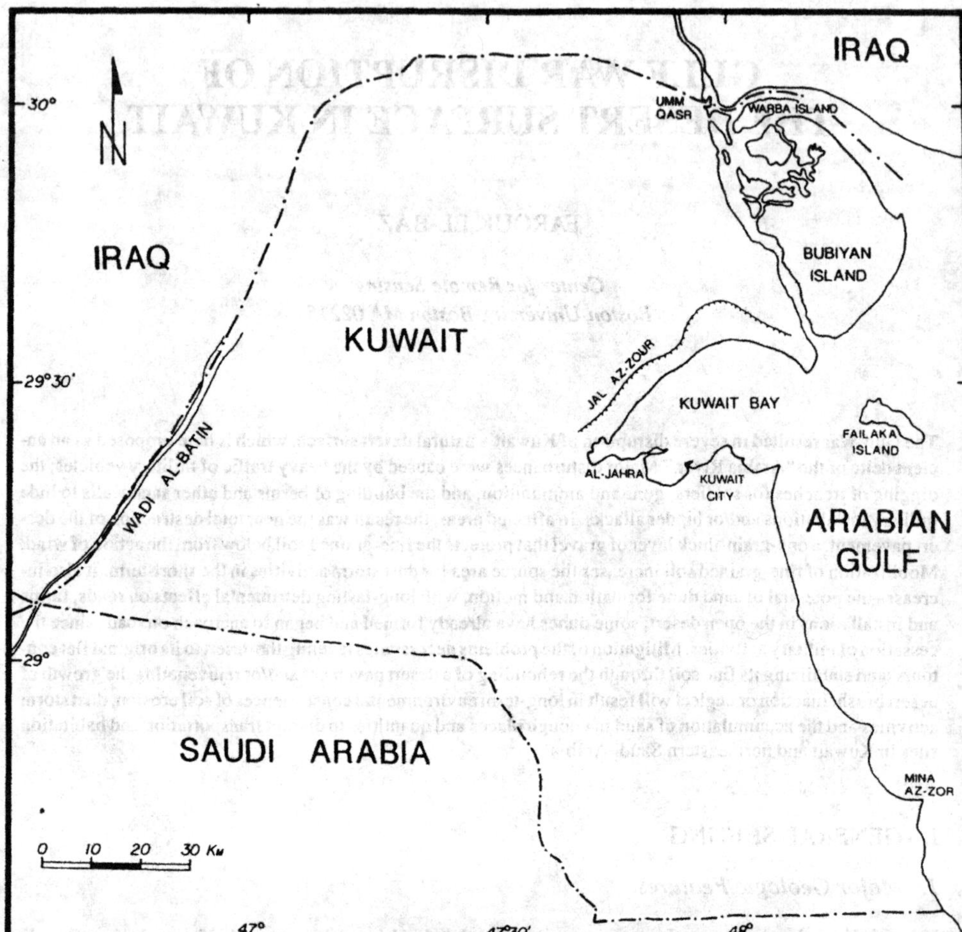

Figure 1. Map of Kuwait showing its major geographic features.

contact with the overlying Miocene and Pliocene-age Fars and Ghar Formations, an indication of sub-aerial exposure and erosion. The Fars and Ghar Formations are primarily calcareous sandstones, sandy limestones, clay and sand. Perhaps as much as half of the deposit is made up of unconsolidated sands. Completing the section above the Fars and Ghar Formations is the Dibdiba Formation, which is divided into an upper and lower members based on grain size. The lower member is a gritty sandstone whereas the upper is a coarser

---

Figure 2. Part of the Arabian Peninsula as viewed by NOAA AVHRR images acquired in September 1990, with the tributaries and partly-buried channel of the "Arabia River" highlighted. Note that western Kuwait is part of the ancient delta of the river. (COLOR PLATE VII)

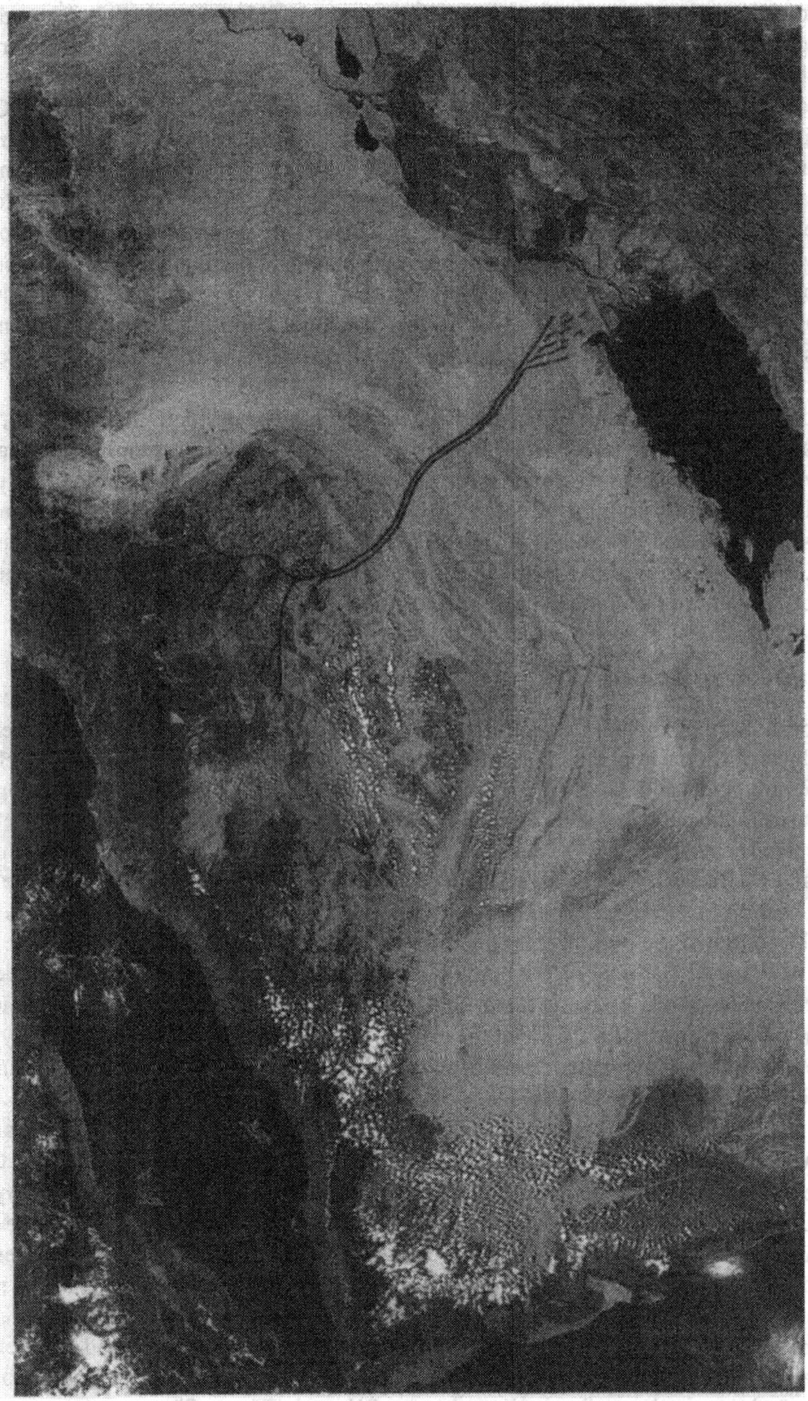

more pebbly sandstone. Quarternary deposits include lag gravels that blanket much of the interior of Kuwait and coastal deposits including unconsolidated marine sands, mudflats and supra-tidal sabkha surfaces (USGS, 1963), most of which originated as sediments from Shatt Al-Arab waterway.

Structurally, Kuwait lies on the Arabian Shield, an area noted for its stability since Cambrian times. The Shield tilts slightly to the northeast, giving rise to sedimentation of the Arabian Shelf consisting of a sequence of laterally extensive, but thin limestones, marls, shales, sandstones and evaporites. North-south trending anticlinal structures developed with gentle warping and uplift, serve as traps for Kuwait's oil reservoirs. In addition to cratonic warping, evidence for salt diapirs exists with magnetic low signatures on the anticlinal structures. Of particular interest is the presence of very strong northeast and northwest trending lineament features which may reflect structural controls. The most striking of these is the Wadi Al-Batin lineament (Figure 2), which continues in a southwest direction into Saudi Arabia.

Topographically, Kuwait is a flat sandy desert that may be divided into two parts. The north is a hard, flat stone desert with shallow depressions and low hills running northeast to southwest and ending near Al-Raudhatain, an area of underground water storage. The principal hills in the north are Jal Az-Zor (145 m) and the Liyah Ridge. The southern region is a treeless plain covered by sand; the Ahmadi Hill (125 m) is the sole exception to the flat terrain. In addition to Wadi Al-Batin, the only other valley of note is Ash-Shaqq, a portion of which lies within the southern part of the country.

2. *Morphologic Characteristics*

The morphology of Kuwait is controlled by the fact that most surface materials appear to be products of fluvial deposits. The author proposes that these are delta deposits which are part of Iraq's rivers and mostly the now dry Arabia River. The latter was active during humid phases in the past. Based on comparisons with nearby regions, the last of the wet climatic episodes most probably ended approximately 5,000 years ago (Solle, 1966; Glennie, 1970; and El-Baz, 1988). The river channel of Wadi Al-Batin probably drained the east central portion of the Hijaz Mountains in western Saudi Arabia, which would have provided the source of the omnipresent gravels (Holm, 1960).

East of Wadi Al-Batin along the borders between Kuwait and Saudi Arabia, the desert surface is made-up of a calcretic plain. This unit is composed of rock debris surrounding hills and gravel ridges. Often, parts of the unit are covered with duricrust, a secondary surface deposit of salts (gypcrete, calcrete, and partly silcrete) that form a crust of saline or alkaline soils, which is further evidence of deltaic origins.

The largest morphologic unit of Kuwait lies east of the calcrete unit and northeast of Wadi Al-Batin. In this unit, the terrain is rugged and is composed mainly of a vegetated sand sheet, locally refered to as Al-Huwaimliyah plain. The ruggedness is caused by the accumulation of sand in the lee side of the wind behind desert shrubs. Spaces between the sand piles are composed of a thick bed of wind-blown sand. This type of terrain results from the coalescence of sand drifts in the lee of desert shrubs; the shrubs act as the main agent of sand stabilization. Therefore, destruction of the vegetation by extended droughts, over-grazing or vehicular overuse would result in the mobilization of the sand in the otherwise stable sand sheet. Several fingers of the gravelly plain extend southwestward into other units as shown in the surface sediment map of Kuwait (Figure 3).

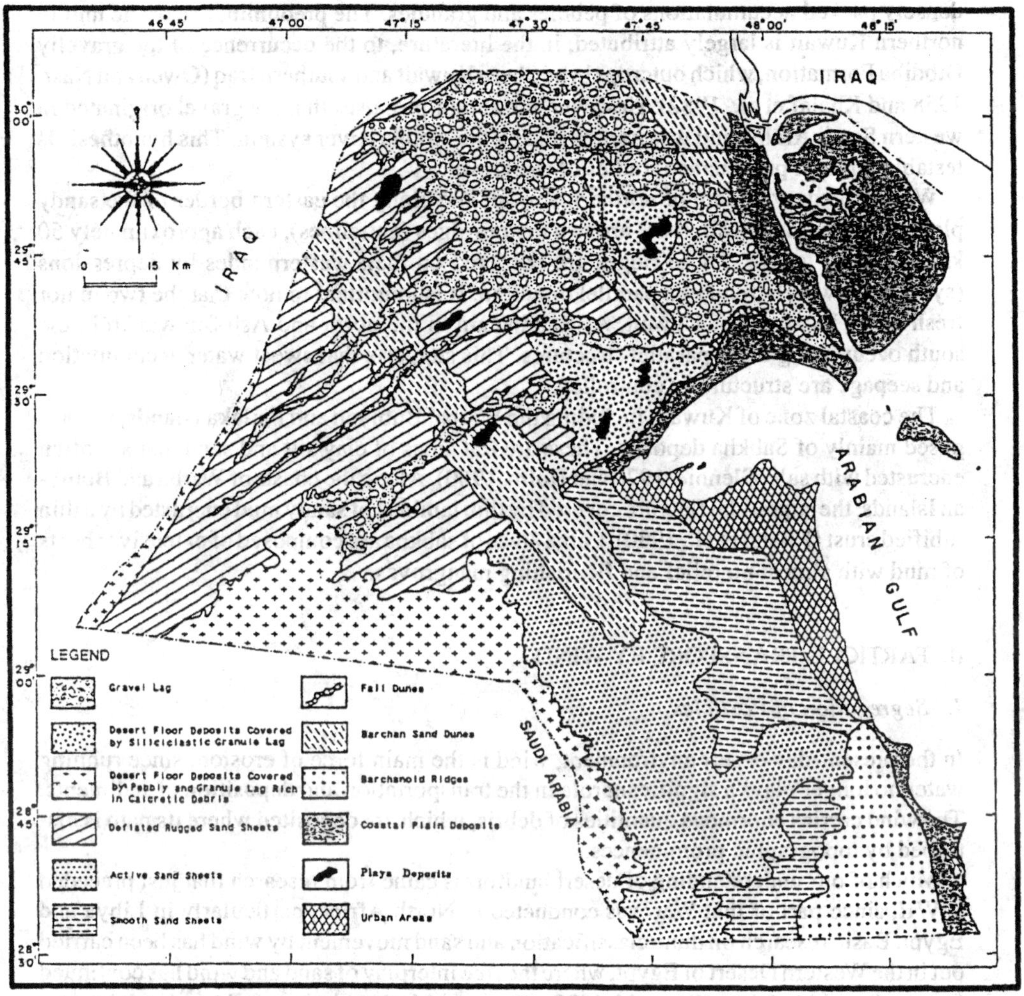

Figure 3. Map of the surface sediments of Kuwait (after Gharib et al., 1985). Note the concentration of playa deposits (black patches) in the northern part of the country. These may represent occasional accumulations of water in the ancient delta of the Arabia River.

In south-central Kuwait the land is composed of a sandy plain (Figure 3) that is basically a mobile sand sheet composed of light-colored sand with a rippled surface. The sand is mostly wind-blown and accumulates in broad depressions (Khalaf et al., 1980). The source of the sand may be partly from the plains of northern Kuwait and southern Iraq, and also from the periodic mobilization of the rugged, vegetated sand sheets of Kuwait.

Northern Kuwait is covered by a vast gravel plain with particles ranging in size from 4 mm to 10 cm in diameter. This unit is considered a desert pavement of lag gravels, which are fragments of rock too large to be moved by the wind and therefore "lag" behind the finer particles that move with the wind. Their concentration varies from scattered pebbles to densely packed accumulations of pebbles and granules. The predominance of the unit in northern Kuwait is largely attributed, in the literature, to the occurrence of the gravelly Dibdiba Formation, which outcrops in northern Kuwait and southern Iraq (Owen and Nasr, 1958 and Khalaf et al., 1982). However, the author proposes that the gravel originated in western Saudi Arabia and was transported by the Arabia River system. This hypothesis is testable through provenance studies of the gravel.

Within the gravel plain of northern Kuwait and along the eastern border of the sandy plain of southern Kuwait, there are two major ridges (anticlines), each approximately 50 kilometers in length. These ridges are bordered on their western sides by depressions (synclines), which trend in a parallel direction. It is important to note that the two major fresh water localities of Kuwait, Al-Raudhatain in the north and Ash-Shuwaykh in the south occur along these parallel structures. This suggests that sweet water accumulation and seepage are structurally controlled.

The coastal zone of Kuwait, including the Warba, Bubiyan and Failaka Islands, is composed mainly of Sabkha deposits. These are flat areas of clay silt and sand that are often encrusted with salt (Glennie, 1970 and Holm, 1960). Along the coasts of Warba and Bubiyan Islands, the supratidal flats are covered by thin laminae of sandy mud encrusted by a thin lithified crust (Khalaf et al., 1980). Large areas of sabkha are composed of extensive sheets of mud with almost no relief and flourishing mangroves.

II. PARTICLE TRANSPORT BY WIND

1. *Segregation of Particles*

In the present-day desert environment, wind is the main force of erosion, since running water no longer plays a significant role in the transportation and deposition of sediments. The wind carries enormous quantities of debris, which are deposited where its path is disturbed by topographic prominences.

Much of our understanding of desert landforms came from research that just preceded WWII, since part of that War was conducted in North Africa, particularly in Libya and Egypt. Basic research on dune classification and sand movement by wind has been carried out in the Western Desert of Egypt, where the free interplay of sand and wind has continued for a long period of time (Bagnold, 1933). Geomorphic descriptions followed by laboratory experiments led to Bagnold's (1941) classic treatise on the physics of sand movement by wind based largely on observations made in the Western Desert of Egypt.

During the past two decades, research on the landforms of deserts and semi-arid lands was intensified, in part from scientific interest in the effects of prolonged droughts in the Sahel belt of North Africa. As a result of this emphasis, several books were published that treat the geomorphology of arid lands (Butzer and Hansen, 1968; Glennie, 1970; Cooke and Warren, 1973; Doehring, 1977; Mabbutt, 1977; El-Baz, 1984; and El-Baz and Hassan, 1986).

As shown by these workers, prolonged wind erosion creates isolated hills in a basically flat terrain. Also, corrasion features shaped solely by the wind can be 100 km in length (Mainguet and Callot, 1978). Yardangs, the smaller wind-sculpted features that resemble inverted boat hulls, occur in numerous localities. In addition, the role of vorticity in developing lineations by wind erosion was established by Whitney (1978). This role eliminates the necessity for shifts in the orientation of ventifacts relative to the surface winds.

A striking characteristic of the desert environment is that wind deposits are not mixed-up and scattered at random. Rather, they are discretely zoned in response to the capacity of the wind to sort out and segregate particles by grain size. The finest particles (clay and silt particles, up to 0.05 mm in size) are winnowed out wherever they are exposed and whirled into the atmosphere as dust, settling out of suspension beyond the zones of high wind energy (Breed et al., 1980). Saharan storms, for example, carry such fine dust particles concentrated at altitudes of 1.5 km and 3.7 km in the atmosphere to be deposited in the western Atlantic Ocean (Prospero and Carson, 1972).

In contrast to the suspension in the atmosphere of the finest particles, fine-to-medium sand saltates readily in strong desert winds. By the saltation process, the wind segregates particles of 0.05 to 0.5 mm sizes from the clays, silts and gravels and shepherds them into dunes. The movement of sand by saltation depends on the nature of the surface. Sand particles jump to higher levels if they saltate on gravel than on a sandy surface. One measurement showed that 90% of sand moved below 87 cm on a gravel surface, whereas the same amount moved below 31 cm on a sandy surface (Bradshaw et al., 1978, p. 217).

Particles too large (0.5 to 2 mm in diameter) to be lifted off the surface by the bombardment of saltating grains may gradually and erratically move or roll along the surface. With high winds the whole surface covered by such grains appears to be creeping slowly along the wind direction.

This process of segregation of particles by the wind into varying sizes usually results in the formation of vast, flat plains that are veneered with a surface of granules or pebbles, larger than 2 mm in diameter, commonly well-sorted and usually one grain thick. This lag forms an armor that stays in equilibrium with the strongest winds. The process of surface creep, by which the wind distributes the coarse materials, seems imperceptible except under rare high-velocity wind conditions. Removal of the coarse lag results in immediate deflation of the suddenly unprotected, underlying silts and sands, which produces lasting scars (Breed et al., 1980).

2. *Dust Storms*

Dust and sand storms in Kuwait are particularly frequent in spring and summer due to strong surface winds, locally called "shamal" from an Arabic word meaning north, which are accompanied by vertical movements of heated air during these two seasons. As discussed earlier in this volume, from April to August each year, large clouds of dust and sand are raised and carried by the dry fresh northwesterly winds. Dust/sand storms, rising dust and suspended dust, occur 13% of the time throughout the year. However, this amount is doubled during daytime hours in April through August (Safar, 1985) with the highest amount of dust in the air occurring in June and July (Figure 4).

For dust to be raised and remain in the air, three conditions must coexist: (1) The land surface must be dry and dusty; (2) the wind should reach at least moderate speeds in order to distribute the surface dust; and (3) the air must be unstable to allow extensive vertical motion. In stable conditions, turbulence caused by the wind is checked and dust rises a few meters. However, strong "lapse rates" would carry the dust to high levels (Safar, 1985). Figure 5 shows the annual variation of dust at the Kuwait International Airport from 1962 to 1984. The author suggests that the increase in dust activity in the 1980's, clearly shown in Figure 5, was due to disturbance of the desert surface during the Iran–Iraq War of 1980 to 1988.

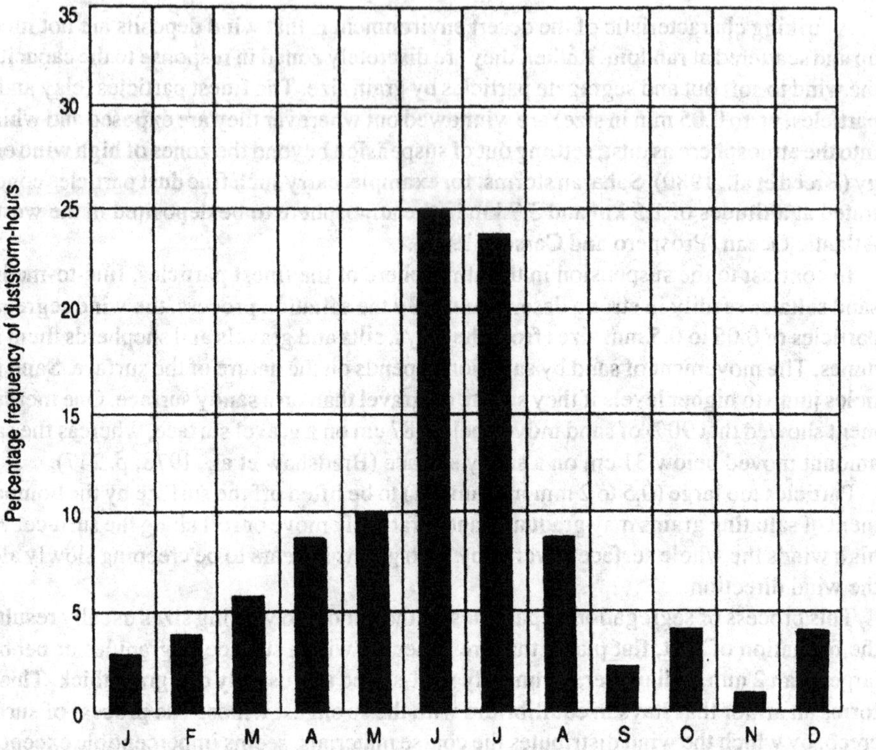

Figure 4. Monthly variation of dust storms at Kuwait Airport during the period 1962–1984 (after Safar, 1985).

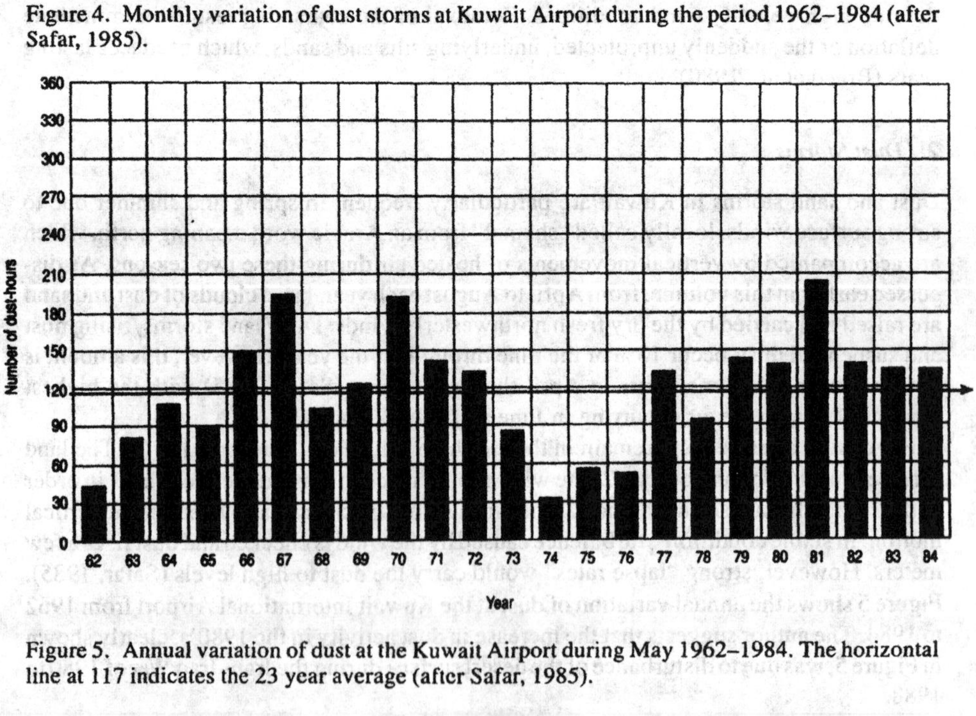

Figure 5. Annual variation of dust at the Kuwait Airport during May 1962–1984. The horizontal line at 117 indicates the 23 year average (after Safar, 1985).

According to Chepil and Woodruff (1957), the concentration of suspended dust in the air at 2 meters above the surface in micrograms per cubic meters is related to the visibility in kilometers. For example, the visibility of 11 km corresponds to a mass concentration of 2,80 0 μg/cubic meter, and the visibility of 1 km to 56,000 μg/cubic meter. The latter is the case during dusty summer days in Kuwait.

As summarized by Safar (1985, p. 20), the types of dust in Kuwait include the following:

- *Dust/sand storms*: These are caused by fresh, strong winds of 18 knots or more. Horizontal visibility is reduced to less than 1,000 meters; when it deteriorates to less than 200 meters, the dust/sand storm is regarded as "severe".
- *Rising dust*: This phenomenon is caused by moderate winds. Horizontal visibility averages better than 1,000 meters.
- *Suspended dust*: Suspension of dust occurs due to light or calm winds, with horizontal visibility reduced to less than 1,000 meters. When the wind is moderate, visibility ranges from 1 to 5 kilometers.
- *Haze*: This phenomenon is caused by very small solid particles of dust, smoke or salt, which reflect impinging solar rays in all directions. Conditions of haze reduces visibility to 5 km or more.

In Kuwait, dust generally remains in the lower part of the atmosphere, near the ground. However, very fine particles may be carried aloft as high as 2,700 meters. Meanwhile, during severe dust storms it can reach 5,500 meters above the ground surface, as reported by pilots of Kuwait Airways (Al-Kulaib, 1990).

3. *Sand Accumulations*

As stated above, sand-sized grains are lifted by the wind and accumulate in deposits in the form of dunes or flat blankets called sand sheets. The sheets may cover enormous areas in Kuwait, but more commonly, they occur as long and narrow sand deposits sometimes without any slip faces, referred to here as streaks. In some places, these streaks occur in straight lines in the lee of topographic obstacles in the way of predominantly northwesterly winds. In others, they create streamlined patterns similar to those described in the Western Desert of Egypt (El-Baz and Maxwell, 1979).

Where patches of fine sand occur without the stabilizing larger grains, the surface becomes rippled with undulations that are transverse to the wind. The wavelength of sand ripples is commonly from 10 cm to 50 cm. The ripple index, which is the ratio between the wavelength to the amplitude, is generally about 20 cm (Sharp, 1963). Ripples may also grow in sands with an armor of coarse granules, up to 4 mm in diameter, with wavelengths of one or two meters (Weir, 1962). These are called granule ripples, which may be truncated by very strong winds and thus create a bumpy surface.

Accumulations of sand take numerous forms, which may be: (1) simple dunes where the individuality of each dune is clear and separable; (2) compound dunes in which two or more of the same type coalesce or combine by overlapping each other; or (3) complex dunes that result from the combination of two different types of dunes.

Dunes may be free, or obstacle-related. All types of free dunes may be active, semi-fixed or fixed by vegetation. The degree of dune fixation depends on the amount of vegetation cover. In general, sand dunes rarely occur as a single unit. More commonly, groups of dunes occur together. Such groupings are described as dune bundles, dune fields and sand seas (El-Baz, 1986).

In general, dunes occur in a variety of shapes. Numerous terms have been employed to classify dune types. In many cases, terms used by local desert dwellers are used to describe the different types. This often results in referring to the same type by various names in different localities. In other classification schemes, a genetic connotation is implied. For example, the term transverse dunes indicates formations normal to the wind direction. Using a purely morphologic classification, at the risk of oversimplification, dune shapes are divided into three types: linear, domical and crescentic (El-Baz, 1986).

Linear dunes are sand accumulations in the form of a long ridge. They are known as "seif" dunes in the Arabian Peninsula because of their likeness to the sword, which in Arabic is called seif, specifically the curved scimitar. In many other deserts these have been described as longitudinal dunes, and in Egypt as Ghurud. Such ridges are usually formed where there is a large supply of sand and a persistently strong wind, such as in the Western Desert of Egypt (Gifford et al., 1979).

Domical dunes are mound-like sand accumulations in the form of a dome, pyramid or star. Some dome dunes occur in Kuwait as gently sloping mounds of circular or elliptical shape that normally show no slip faces. However, some may show a small slip face suggesting the start of a crescent-shaped dune. It is generally believed that dome dunes occur in a strong, multi-directional wind regime.

Crescentic dunes are the most common in Kuwait with the crescent-shaped or barchan dunes having arms pointing downwind, and U-shaped or parabolic dunes with arms pointing upwind. They range from simple shapes to coalesced crescents to crescentic or barchanoid ridges. These dunes occur where there is a limited sand supply and more of an energetic wind than in the case of linear dunes.

Fields of crescentic dunes occur in Umm Negga and Al-Huwaimiliyah regions of Kuwait. Smaller accumulations exist in Umm Al-Aish and Al-Atrah regions. At Umm Negga, their average width is 170 m and their average height is 8 meters. Those of Al-Huwaimiliyah are smaller, generally 20 m wide with an average height of 2 meters.

Comparison between the textural characteristics of the major dune fields of Kuwait and those in southern Iraq shows that Al-Huwaimiliyah dune sands are texturally similar to Al-Nasiriyah dune sands, while Umm Negga dunes are similar to those of Basra (Foda et al., 1984). Furthermore, satellite data confirm the extension of the two main parallel dune belts into southern Iraq. The first, extends from Najaf to Al-Huwaimiliyah passing through Al-Nasiriyah in a southeasterly direction. The second, extends from south of Basra to Umm Negga region of Kuwait. Figure 6 shows the relationship between the dunes of Kuwait and those in southern Iraq.

In addition to distinct dune forms, sand accumulations occur as "obstacle deposits". These occur because topographic prominences in desert surfaces nearly always affect the wind pattern and invariably cause the eolian deposition of sand. A few large topographic prominences such as mountains cause major variations in the wind flow. These variations result in the formation of different types of dunes windward, leeward and on either side of a topographic barrier (Manent and El-Baz, 1980). Small scale topographic rises are just as effective in causing the accumulation of sand into numerous forms (Mabbutt, 1977). Since the mobilization of sand due to the Gulf War, a variety of these have been observed in the desert of Kuwait, including the following:

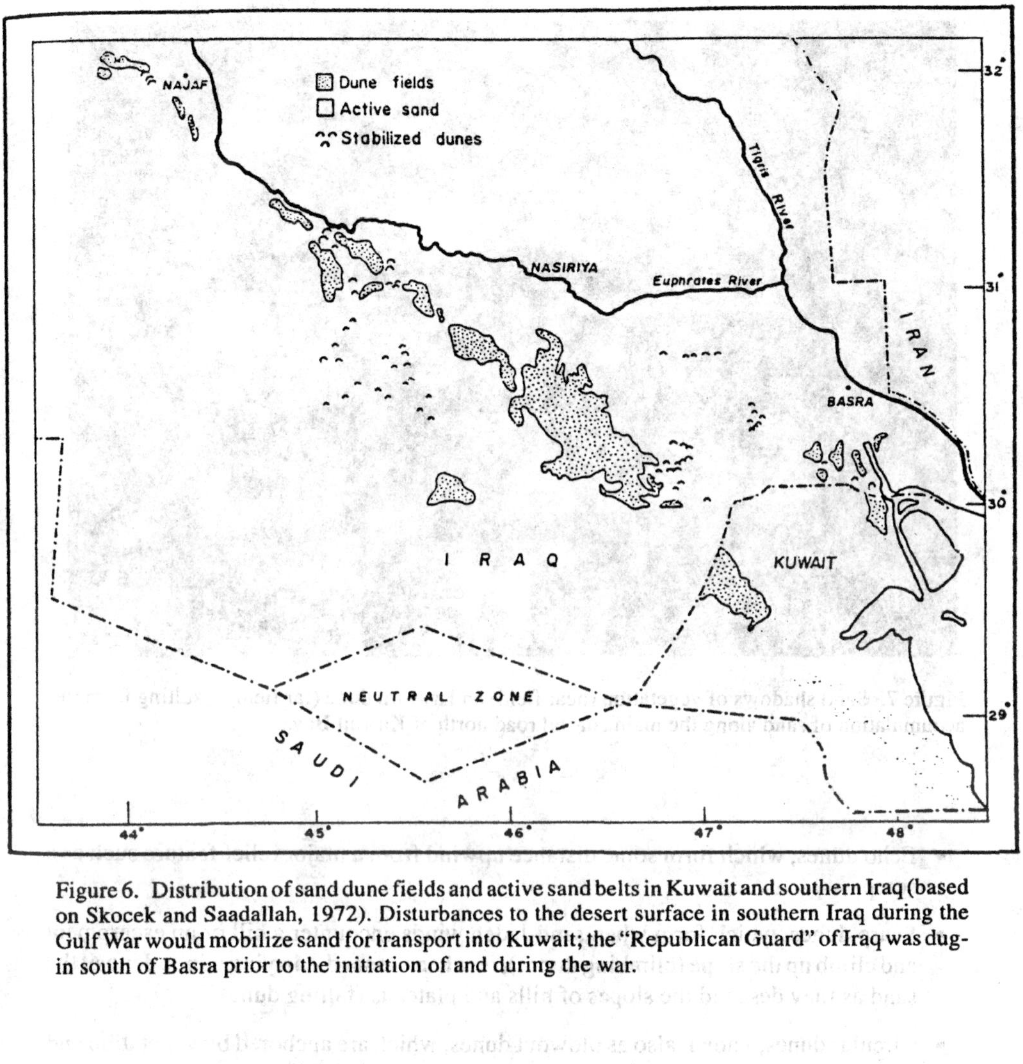

Figure 6. Distribution of sand dune fields and active sand belts in Kuwait and southern Iraq (based on Skocek and Saadallah, 1972). Disturbances to the desert surface in southern Iraq during the Gulf War would mobilize sand for transport into Kuwait; the "Republican Guard" of Iraq was dug-in south of Basra prior to the initiation of and during the war.

- Sand shadows, which are elongate deposits in the lee of an obstacle such as a rock, hill, or clump of vegetation (Figure 7).
- Sand drifts, which are elliptical mounds less tapered than sand shadows, resulting from the deposition in gaps between obstacles or from vertical separation of flow in the lee of steep slopes.
- Lee dunes, which result from the buildup and downwind extension of sand shadows and sand drifts.
- Windward dunes, which result from the deposition of sand upwind from an obstacle such as a hill or building.

Figure 7. Sand shadows of vegetation (near field) and a scarp dune (far field) resulting from the accumulation of sand along the main coastal road north of Kuwait Bay.

- Echo dunes, which form some distance upwind from a major relief feature such as a scarp.
- Scarp dunes, which form when sand-laden winds encounter a hill or an escarpment and climb up the slope (climbing dunes), or when the sand-carrying winds deposit the sand as they descend the slopes of hills and plateaus (falling dune).
- Circular dunes, known also as blowout dunes, which are anchored by vegetation and may have slip faces sloping in one or many directions depending on what parts of their rims are not stabilized and therefore free to move.

4. *Desert Pavement*

Desert pavement is a natural residual concentration of wind-polished, closely packed pebbles, boulders and other rock fragments making a desert surface (Figure 8). In a mature desert surface, wind action and sheetwash from occasional rainstorms have removed all smaller particles. The remaining one-grain-thick layer of rock fragments protects the underlying fine-grained material from deflation by the wind. In some cases, the fragments are cemented by mineral matter and form a crust. Figure 9 shows the process by which a desert pavement is formed.

Figure 8. Desert pavement in Kuwait (top; match box in upper left shows scale) as a one-grain-thick layer of pebbles on top of fine grained soil (bottom). (COLOR PLATE VIII)

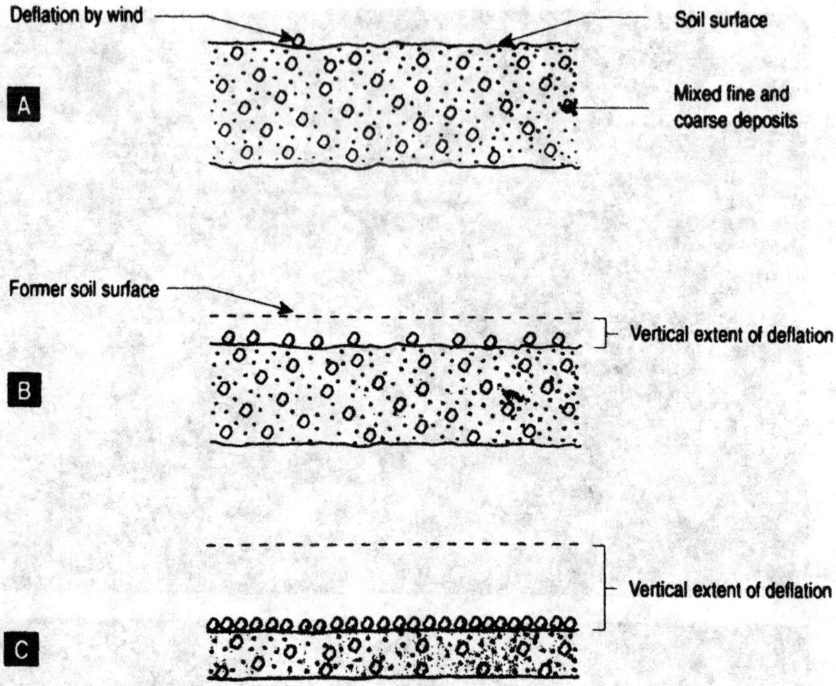

Figure 9. Stages of the development of a gravel sheet as established by research work at the Kuwait Institute for Scientific Research (KISR).

Most of the desert surface in Kuwait is covered by lag particles. The nature and occurrence of this lag cover can be defined by texture (size and sorting), concentration, and composition. Particle size of the lag cover ranges from fine granules to large boulders. However, it generally has a bimodal size distribution of gravel (an unconsolidated natural accumulation of rounded rock fragments consisting predominantly of particles larger than sand with diameters greater than 2 mm to over 10 cm, such as boulders, cobbles, pebbles, and granules or any combinations of these) and pebbles (small, roundish, usually water-worn stones having a diameter of 4 to 64 mm).

Most of the northern desert of Kuwait is covered by desert pavement. The relative abundance of the lag deposits in this area is ascribed by the author to be due to the fact that the region is part of the ancient delta of the Arabia River. The desert pavement occurs as lag sheets and lags associated with ridges. Lag sheets form an extensive blanket of gravel in the northern desert and can be categorized into three subclasses: granule lag sheets, pebbly lag sheets, and mixed sheets of granules and pebbles.

In general, lag size increases towards the western parts of the desert, i.e., the size of gravel in the central parts of the desert ranges between 2–4 cm, while at the far western parts fragments, in the order of 10 cm, are found. The author suggests that this is most likely due to the fact that the gravels were transported from the highlands of central Arabia via Wadi

Al-Batin; deposits closer to the source would contain larger particles, and those farther away from the source would have finer particles.

From the characteristics of the desert pavement of Kuwait (Holden, 1991), it is clear that gravels were first moved by water currents (Al-Asfoor, 1980) and deposited in a sheet together with finer fractions, which were later blown away by deflation. As a result of the continuous deflation, pebbles and granules were concentrated on the surface until a deflation armor was acquired (Khalaf et al., 1980). The gravel ridges were formed by reversed topography (Khalaf et al., 1982) as a result of the preferential deflation of flood plains dissected by fluvial channels in the fan-shaped delta of the Arabia River.

III. PRE-INVASION CONDITIONS

Developing land for agriculture has increased in Kuwait during the past two decades. This increase in agricultural activity resulted in the disruption of the desert pavement in developed areas and the exposure of soil to wind action. This, in turn, has increased the movement of dust and sand, although the affected areas were relatively small.

Moreover, drastic increases in dust and sand storms in the northern Gulf region were noticed in the 1970's due to the vast increase of agricultural expansion, as part of major desert reclamation projects, in the Western Desert of Iraq. Satellite images show the initiation of dust storms in southwestern Iraq to spread over the eastern part of the Arabian Peninsula, as discussed earlier in this volume.

Similarly, growth of urban communities results in the disturbance of the stable desert surface. Most of the enviromental effects of urbanization occur during the digging in the ground, where soils are exposed to eolian erosion. Such effects were particularly obvious during the expansion of Al-Jahrah, west of Kuwait City.

Further inland, disruption of the desert surface from the oil industry resulted from exploration and development activities, including geophysical surveys, drilling operations, building of roads and laying of pipelines. These activities usually require a large number of vehicles, which criss-cross the desert surface affecting both the desert pavement and the scant desert vegetation.

As an example, the aforementioned morphologic unit of Al-Huwaimliyah plain, the vegetated sand sheet, extends into the Burgan oil field in south central Kuwait. This land area was badly degraded due to excessive vehicular activities and the desert surface was disturbed. Because disturbance of this type mobilizes the sand, vehicular traffic in the open desert in Burgan field has been prohibited during the past six years. Due to this prohibition, the region began to flourish and the sand sheet was once again stabilized.

Military installations in the open desert also have had some impact on the environment of Kuwait's desert. Military bases have greatly suffered from the effects of destabilizing the desert surface in terms of having to constantly alleviate the effects of sand accumulations around buildings and their entrances. The study of sand deposits within Ali Al-Salem and Ahmad Al-Jahbir bases provided insights into accumulations on roads (Gharib et al., 1985). Three cases of pavement profiles on a slightly raised bed were considered: a flat pavement; one where the underlying bed was slightly elevated on the upwind side of the road; and another with the elevation on the down-wind side of the road (Figure 10).

In all three cases sand accumulated on either side of the raised road bed, regardless of its slight elevation, a road bed represents a topographic impediment to the wind. In the case of the flat road bed and pavement, sand accumulated on the down-wind side of the road. The

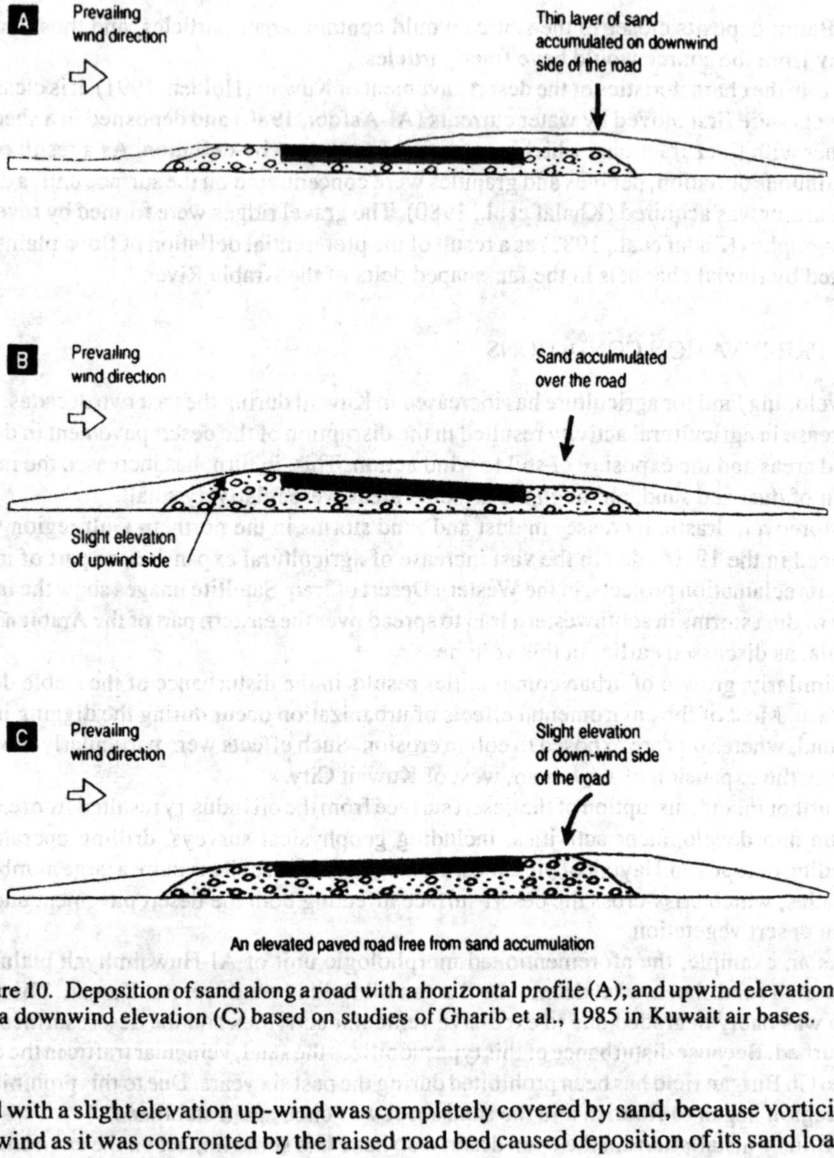

Figure 10. Deposition of sand along a road with a horizontal profile (A); and upwind elevation (B); and a downwind elevation (C) based on studies of Gharib et al., 1985 in Kuwait air bases.

road with a slight elevation up-wind was completely covered by sand, because vorticity of the wind as it was confronted by the raised road bed caused deposition of its sand load. In the case of the slight elevation of the road bed and pavement on the down-wind side, the road remained free of sand. In this case, the wind energy was increased as it moved upslope and carried the sand with it across the road to be deposited beyond the down-wind elevation of the road bed.

In the future, similar precautions and protective measures will have to be taken to safeguard installations in the open desert such as roads, airport runways, helicopter pads, oil wells and storage tanks, checkpoints, and residential sites in the oil fields of Kuwait. Each site will have to be considered separately, because remedies may differ due to local topographic characteristics, nature of the desert surface, and the potential of particle transport by the wind.

IV. DAMAGE ASSESSMENT

1. Vehicle Tracks and Roads

The movement of the great numbers of tanks and mechanized vehicles on the desert surface of Kuwait has adversely impacted the desert pavement, particularly in the western part of the country. During the invasion, columns of Iraqi military equipment forced their way along parallel lines both inside and just west of the border of Kuwait. Such heavy equipment forces the surface cover of pebbles and cobbles deep into the sand and excavates large amounts of soil on either side of the tracks.

The net result of vehicle movement on a naturally packed surface is the decrease in the density of packing of sand grains (El-Baz, 1979). This phenomenon increases the potential of grain movement by the wind, resulting in the mobilization of previously protected fine particles. The spread of newly exposed soil on either side of the tracks allows for their detection on satellite images, even if the resolution of the images is less than the width of the tracks (Figure 11).

During occupation, most of the troop movements were along paved roads. However, some areas were trampled by military vehicles causing damage to the desert pavement. This is particularly true in the southwestern and northeastern part of the country as Iraqi soldiers raised berms and other defensive military fortifications. In areas covered by a sand sheet, such as in southeastern Kuwait, the tracks of military vehicles have already been covered by drifting sand. On the other hand, areas covered by a desert pavement remained deeply scarred, such as north of Jal Az-Zor.

2. Trenches and Pits

The Iraqi armed forces spent much of their time during the fall of 1990 digging trenches in the desert of Kuwait (Figure 12). This is particularly true along Kuwait's coastline. Beyond the trenches, pits were dug as hiding places for military equipment and storage areas of ammunition.

After the passage of a shamal season, most pits and trenches have already been filled by drifting sand. It must be cautioned, however, that ammunition pits that are presently covered by sand represent a potential danger in the future. These may be located in the future through the use of ground penetrating radar (GPR) to be excavated and destroyed. Such procedures would again mobilize sand for downwind movement if the dug-up parts are not stabilized by spreading pebbles.

3. Berms and Sand Walls

Berms were built in the Kuwaiti desert to slow down the advance of liberation forces. The major berms were constructed in the south along the Saudi border, in the central part south of Burgan field, and in the northern part of Kuwait, north and west of Jal Az-Zor.

The berm building activity continued throughout the occupation phase. Iraqi military strategists knew that the sand walls would be eroded by the consistent wind. Therefore, berms were sprayed with oil by pipelines that were laid specifically for this purpose from Kuwait's nearby oil fields. These berms represent a source of shifting sand in the long term. At the present time, the soil in these berms is stabilized by the oil crust (Figure 13). Howev-

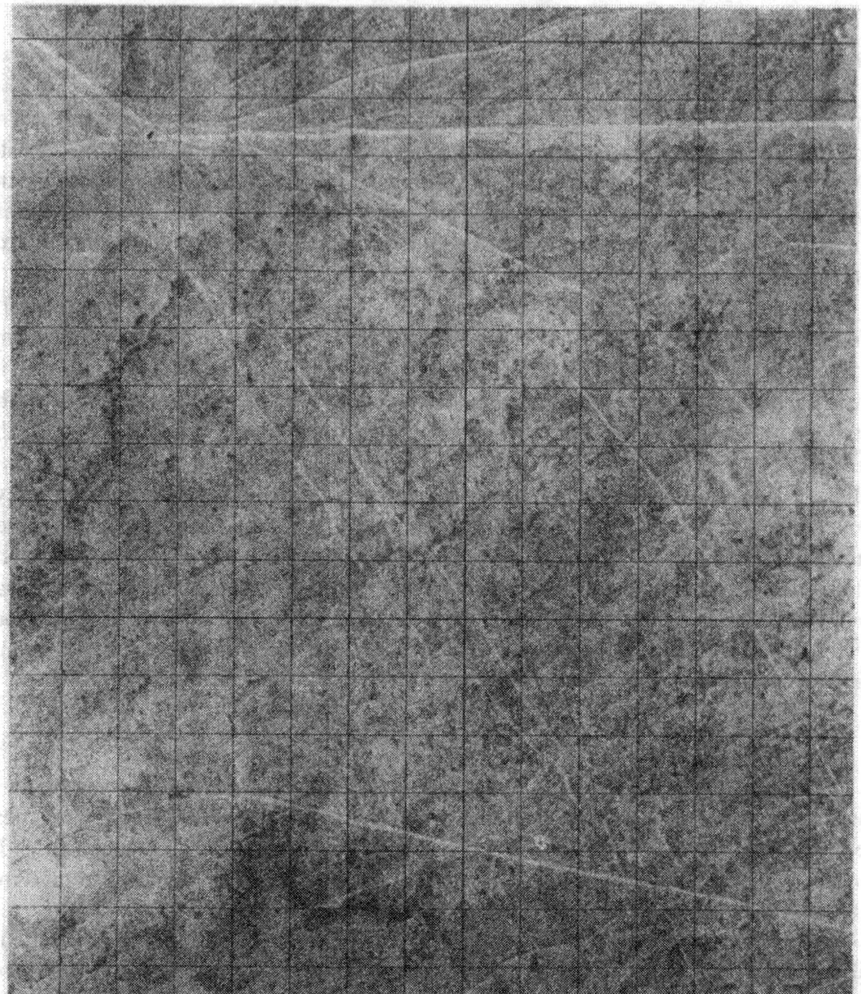

Figure 11. Tracks of vehicles imprinted on the desert surface and depicted on a Landsat Thematic Mapper image of Kuwait's southwestern corner, just east of Wadi Al-Batin.

er, this crust will gradually be dismantled by the wind and fine-grained soil will be regularly released.

Effects of such berms on the spread of sand on the desert surface can be exemplified by the case of the sand wall in Saudi Arabia just south of its border with Kuwait. The latter was constructed to stop potential smugglers. The berm is less than three meters high and a few meters wide at the base. However, its sand was mobilized and spread over such a large area to make it clearly visible on satellite images of 30 meter resolution because the sand was blown down wind, up to one kilometer from the wall (Figure 14).

Figure 12. A trench dug in the desert surface north of Jal Az-Zor that was filled by sand in less than a year from the end of the Gulf War.

In addition to the berms built by Iraqi military personnel, a major berm was constructed during the summer of 1992 by Kuwait's security forces in the northern part of the country to stop a wave of smugglers and political agitators from Iraq following the cessation of hostilities. Since wind erodes the fine-grained material of the berm, it is constantly maintained by bulldozers. This berm will serve as a source of drifting sand and must be monitored as well.

Figure 13. A berm built by Iraqi forces in northern Kuwait and sprayed by crude oil. Behind the berm is an oil-filled trench; the whole area would have been set on fire during an advance of Coalition Forces (Photograph by Jassim Al-Hassan).

4. *Mines and Ordnance Detonation*

As stated earlier, the Iraqi forces have planted between 500,000 and 2,500,000 mines in the desert and coastal areas of Kuwait. The mining was done during the occupation phase to hamper the expected liberation of Kuwait by military action.

Numerous defense lines were established particularly along the border of Saudi Arabia and the coastline of the Arabian Gulf. In addition, mines were laid in areas of high troop concentrations, particularly the tri-border area near the western corner of Kuwait. A map of the mined region was made by security forces of Kuwait following the liberation at the scale of 1:300,000. It is shown as a reduced version in Figure 15.

The laying of mines is a process which results in the disturbance of stabilized soil and exposure of fine grained material to the action of wind. This is particularly obvious in a mine field along the northern coast of Kuwait Bay. In this case, the mines were laid in a soil that was stabilized by a salt crust. The process of digging the land to plant the mines destroyed the drier crust, resulting in the deflation of the soil layer and exposing the mines (Figure 16).

During liberation, several passages were made through the mine field along the southern border of Kuwait. Breakthroughs were made by exploding the mines in groups. This process also resulted in disturbing additional stretches of the desert pavement. In addition, bombing of Iraqi defensive positions by Coalition Forces resulted in cratering of the desert

Figure 14. Effects of wind erosion on a berm in Saudi Arabia just south of its border with Kuwait. The berm is on the sharp northern edge of the white zone; the wind blowing to the southeast caused the jagged southern boundary due to sand transport downwind. Part of a SPOT image acquired on 25 March 1992.

surface, where bombs missed their targets. Ejecta from the craters exposed much fine-grained soil to the wind. However, because of the accurate targeting, these craters are few and far between.

Much of the disruption of the desert surface in the minefields occurred during the post-liberation phase as a result of mine detection and detonation. Crews of mine detectors were followed by mine sweepers and/or detonation experts. The raking of the land and the explosion of mines and undetonated cluster bombs and other ordnance, further disturbed the desert surface.

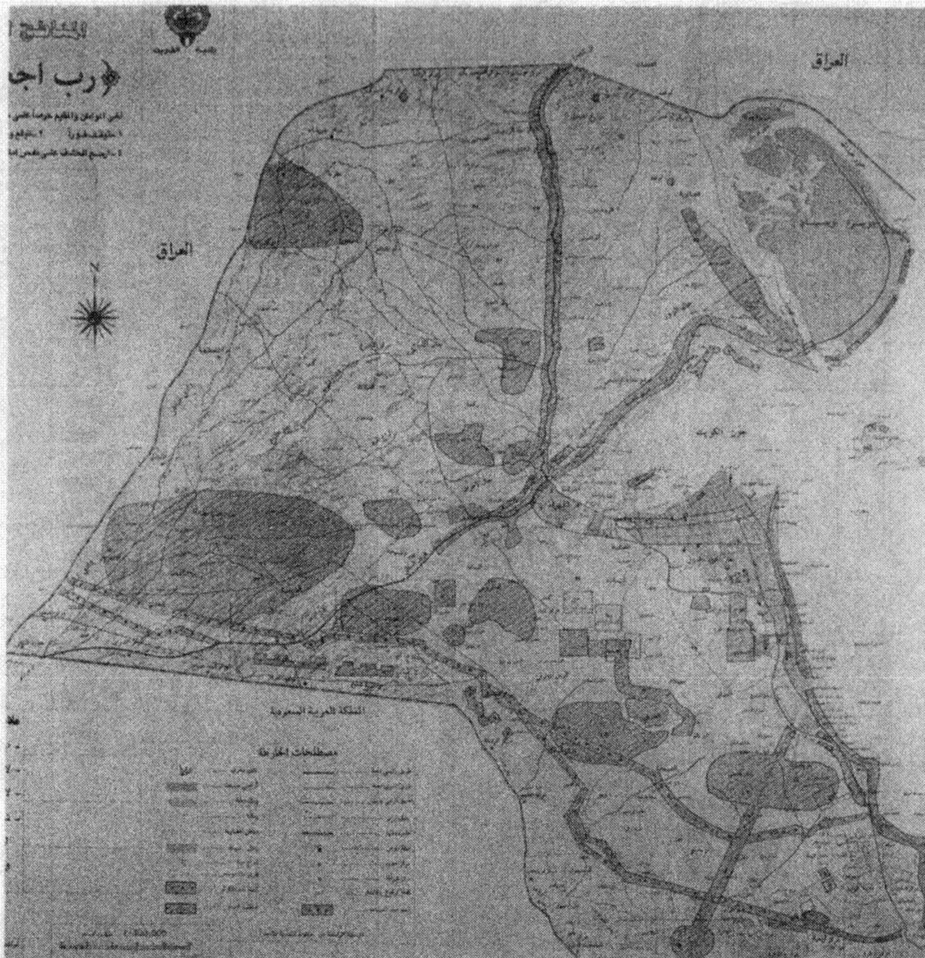

Figure 15. Map of areas of the desert surface of Kuwait that includes mines and unexplored ordnance following the Gulf War as designated by the security forces of Kuwait.

5. *Oil Crust*

As shown in Figure 17, the deposit of oil and soot particles from the plumes of oil well fires are clearly visible on satellite images. When observed in the field, this deposit forms a hardened crust on top of the desert surface. It varies in thickness with an average of 6 cm; the closer to the center of the deposit, the thicker the crust. This is due to the fact that minor variations in the exact orientation of shamal winds caused the spread of the oil rain downwind of the well fires. Naturally, the thicker the deposit, the darker the surface. Minor variations in the spectral reflectance of such deposits are easily depicted on Landsat Thematic

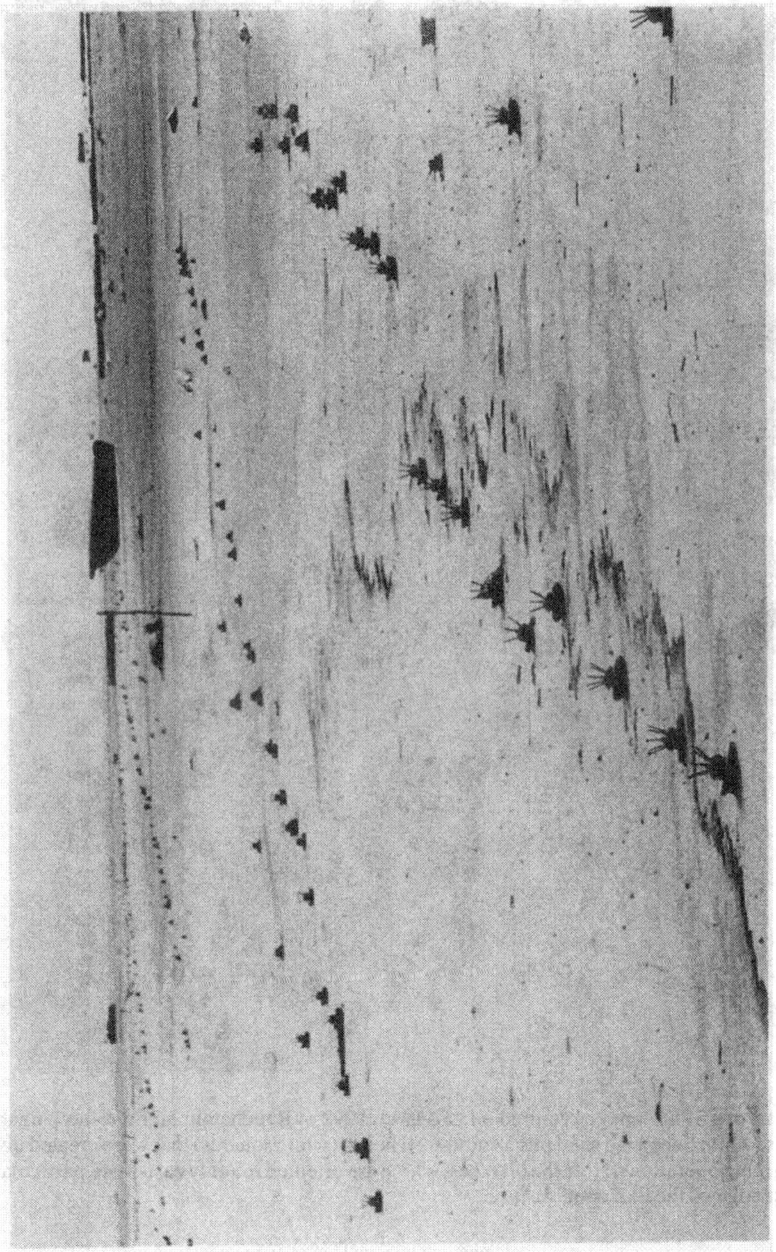

Figure 16. Mines exposed by wind erosion of sand in the intertidal zone along the coast of Kuwait. The mine field was supposed to prevent a marine landing of Coalition Forces (photographed by Jassim Al-Hassan).

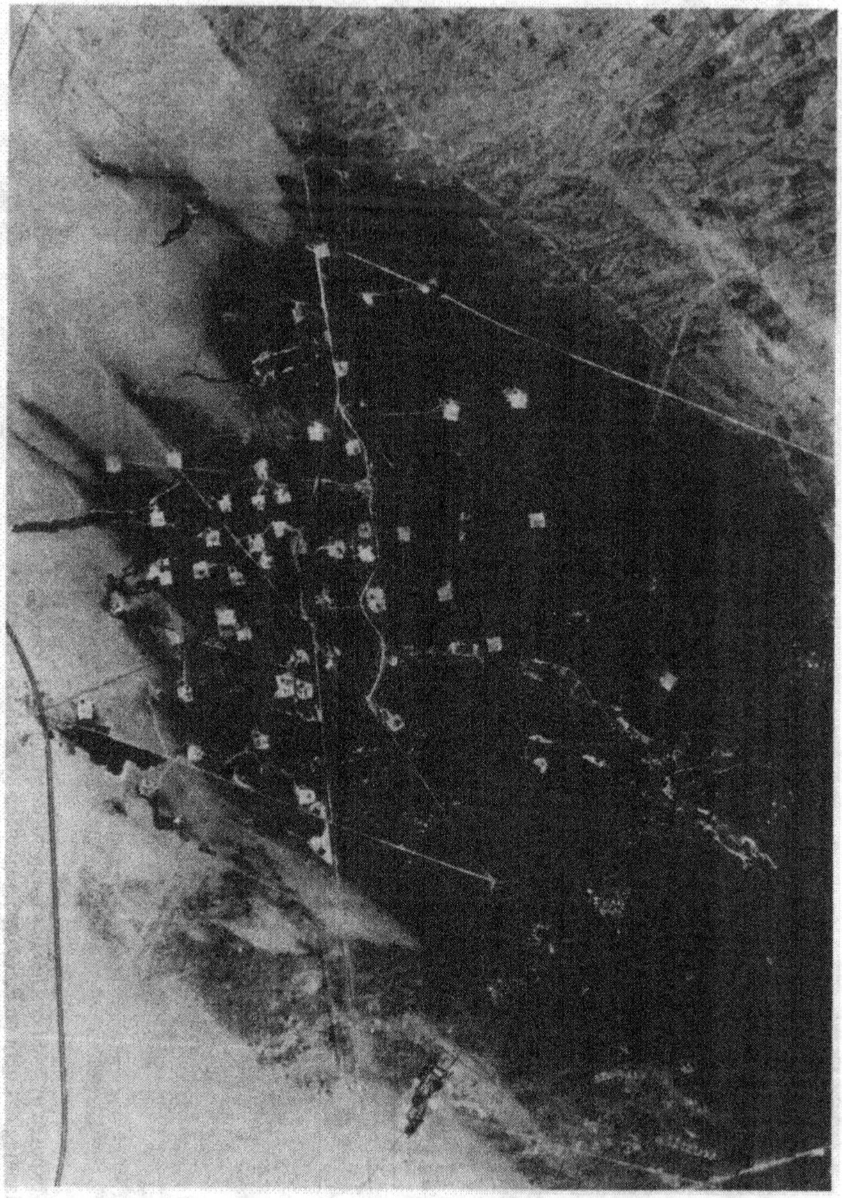

Figure 17. Part of a SPOT image acquired on 25 March 1992 of Raudhatain oil field showing the new dirt roads (white lines) and sand pits around well heads (white squares) that were created during the firefighting operations. Note the oil lakes, which are oriented roughly east-west, particularly on the west edge of the darkened surface.

Mapper images, and it is possible to correlate the reflectance values to the thickness of the oil crust. As it represents a new rock formation, this deposit is here named "tarcrete" and defined as a conglomerate consisting of surface sand and gravel cemented together into a hard mass by petroleum droplets and soot.

The total surface area of Kuwait that is covered by tarcrete is approximately 943 square kilometers. Regarding the problem of desert surface disturbance, this surface area has been basically stabilized. Petroleum sprays have been used in the past to stabilize dune sands (Gore, 1980). Thus, from the sand mobility point of view, the tarcrete has had a positive impact.

The effects of the tarcrete on plant life must be considered separately. Some plants have already died from the oil spray. Other species have flourished, but may not be palatable to grazing animals. Furthermore, the salinity of the soil in areas covered by the tarcrete might change with time, as sulphur compounds react with moisture to increase soil acidity. For this reason, monitoring of the pH values and other soil characteristics is recommended below.

6. *Oil Lakes*

Production of accurate maps of oil lakes necessitated the study of complete coverage of Kuwait by both Landsat Thematic Mapper (TM) and SPOT data as well as maps that are based on field observations. As stated earlier in this volume, change detection techniques were utilized to study the characteristics of oil lakes in TM images. Also, GIS correlations of all data sets allowed checking their accuracy and confirming the produced lake boundries. It was concluded that after a full understanding of the spectral characteristics of the oil lakes in TM images, their accurate mapping could be improved by using the higher resolution SPOT data.

Panchromatic images of SPOT covering all the major oil fields of Kuwait were processed and 1:100,000 scale prints were made. The oil lakes were mapped from these using back lighted prints. The maps were doubled in scale to correlate with other data at 1:50,000 scale. If oil patches that are smaller than one square kilometer are disregarded, the surface area covered by oil lakes in the desert of Kuwait was 35.4 square kilometers in the fall of 1991, as mapped within the Kuwait Oil Company (KOC) operational area by personnel of the Kuwait Institute for Scientific Research (KISR), as shown by the example in Figure 18. The total area of oil lakes was 27.5 square kilometers as mapped from SPOT images obtained on 25 March 1992. The difference between the two figures is easily ascribed to the shrinkage of the lakes due to evaporation of aromatic compounds and the seepage of some of the oil into the underlying soil during the intervening months. The larger figure must be used in the estimation of remediation needs because all oil-saturated soil must be treated or removed, replaced and stabilized.

From the study of satellite images, field observations made during four visits by the author to Kuwait since the end of the war (including helicopter flights), it is estimated that over 30% of the land area of Kuwait has been adversley affected by the war (Table I). As discussed below, some of the resulting problems must be remedied to avert harmful effects in the future.

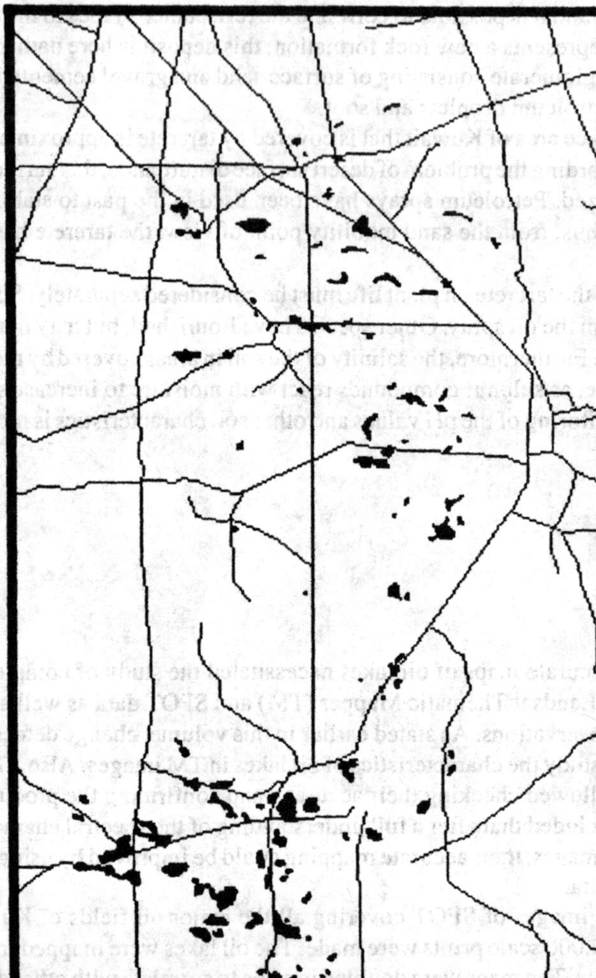

Figure 18. Distribution map of oil lakes in the oil fields of Magwa, Ahmadi and Burgan produced by combining roads from topographic maps and maps of oil lakes by the Kuwait Institute for Scientific Research (KISR) utilizing Geographic Information System (GIS) methodologies.

Table I. Estimate of Total Surface Area of Kuwait Impacted by the Gulf War

	Causes of Disruption	km^2
1	Vehicle tracks and roads	1.1
	• New roads in the oil fields (186.7 km × 6 m) (desert roads included in mine fields)	
2	Trenches and pits	
	• Trenches along the coast and behind berms	0.4
	• Pits for hiding military hardware or personnel	12.0
	• Sand pits around well heads in the oil fields	8.1
3	Berms and sand walls	
	• Length of 4 major berms and shorter segments (each berm affected nearly 1km downwind)	415.0
4	Mine fields	
	• Areas affected by mines and unexploded ordnance, including roads and vehicle tracks in the open desert	3,530.0
5	Oil crust	
	• Surface area covered by oil and soot from fire plumes	943.0
6	Oil lakes	
	• Total surface area affected by oil seepage	49.1
7	Death of plants	
	• Area of desert surface where plants were affected in addition to that impacted by mine fields	500.0
	TOTAL SURFACE AREA AFFECTED BY THE GULF WAR	5,458.7
	PERCENT OF TOTAL SURFACE AREA OF KUWAIT (17,818km^2)	30.6%

V. REMEDIATION AND MONITORING NEEDS

1. *Remediation Approaches*

It is clear from the above that the Gulf War has had a severe effect on the desert surface of Kuwait and neighboring areas. Results of the damage to the desert surface will have short- and long-term effects.

Short-term effects

- Evaporation of aromatic compounds such as benzine from the surfaces of the oil lakes that formed in nearly all of the oil fields of Kuwait. These hydrocarbons may create a hazard to humans, animals and plants.
- Increase of dust storm activity due to the release of fine silt and clay in areas previously protected by course desert pavement. Dust storms create a hazard to aviation in the whole Gulf region. In addition, mineral particles in the air constitute a hazard to humans if lodged in the throat, lungs or eyes.
- Movement of sand and its accumulation over mine fields and ordnance pits, which would hamper the detection of mines and unexploded bombs. In early 1992 several explosions have already caused severe injury to people and death to numerous camels and goats.
- Severe reduction of the plant cover in the open desert due to movements of heavy military vehicles as well as the oil deposits that emanated from the well fire plumes.

Long-term effects

- Increase of sand storm activity due to the mobilization of sand-sized particles. These storms interfere with all modes of transportation and cause road accidents.
- Formation and motion of sand dunes, which encroach on roads, farms, oil installations and settlements in the desert. Once a dune is formed it continues to move for as long as there is wind. For this reason, sand dune encroachment may represent the longest-term impact on the desert surface of Kuwait.
- Increasing soil acidity due to the interaction of moisture with remnants of the oil on the desert surface. This increase would hamper the rejuvenation of natural plants.
- Potential pollution of the groundwater resources from seepage through fractures of oil from the stagnating pools of crude petroleum. This would represent a long-term impact particularly since agriculture in Kuwait depends solely on groundwater reservoirs.

For these reasons, mitigation of the effects of the Gulf War on the desert surface of Kuwait must begin as soon as possible, and continue until the completion of all the steps that could be taken to remedy the problems. Remediation approaches that must be implemented in a timely manner include the following:

a. Filling of empty trenches using the recently mobilized sand to limit the availability of sand-sized particles to the action of wind.

b. Leveling of berms and other sand barriers and walls, because these structures represent a source of sand for future activation and mobilization by the wind.

c. Stabilizing the flattened surface either by an increase of desert plant cover or, more efficiently, by spreading gravel on top of the soil to mimic the natural desert pavement.

d. Pumping of the oil from the lakes and its storage in tanks for future use or mixing with crude.

e. Removing the oil-saturated soil from areas covered by oil lakes to limit potential long-term effects on soil acidity or groundwater pollution.

f. Replacing the removed soil with clean sand to return the land to the original flat contours.

g. Stabilizing the new surfaces by spreading gravel to form an armor against the erosive action of the wind.

h. Seeding of areas of the desert that lost vegetation cover using, as much as possible, seeds of natural plants.

2. *Future Monitoring*

The sources of environmental impacts on desert surface due to the Gulf War are related to specific events including: (1) the invasion of Kuwait by Iraqi forces; (2) the activities that persisted during the occupation phase; (3) the initiation of liberating Kuwait by Coalition Forces and the events that accompanied the retreat of Iraqi troops; and (4) the environmental impact of fighting the oil well fires and the spread of oil lakes on the desert surface of Kuwait.

These impacts can be monitored by the use of satellite images, including: (1) NOAA satellites to monitor the initiation of dust storms; (2) Landsat Thematic Mapper (TM) images to monitor sand accumulation and movement; and (3) SPOT images to monitor changes to the soil. Interpretations of satellite image data need to be checked and confirmed in the field at specific locations. At these stations, instruments should be installed to measure the following:

Dust storms: The exposure of vast amounts of soil to the action of wind causes an increase in the frequency and ferocity of dust storms. These hamper visibility for car and helicopter transport, and cause respiratory ailments to people. The need exists to monitor the initiation of dust storms upwind for advance warning of impending danger.

Sand accumulation: The wind carries enormous amounts of sand in the flat terrain of Kuwait. The sand begins to accumulate where the wind regime is affected by topographic impediments. Such accumulations hamper transport and the operation in the oil fields and other installations in the desert. Thus, a monitoring program needs to be installed to alert concerned personnel for the alleviation of the impact of encroaching sand dunes.

Soil rejuvenation: The oil rain has caused much damage to the natural desert soils in Kuwait. It is feared that additional harm to the soil might be caused by an increase in acidity. A program needs to be instituted to monitor soil conditions and characteristics to allow a remedy of the potential problem in a timely manner.

Plant Growth: The degradation of the soil has caused much damage to the desert plants of Kuwait. Some plants have died in areas covered by the oil rain and other species may appear that would not be palatable to grazing animals. The growth of plants should be monitored in view of the drastic changes in the desert environment.

REFERENCES

Al-Kulaib, A. A. (1990) Weather and Climate in the State of Kuwait (in Arabic). Meteorological Department of the Directorate General of Civil Aviation, Kuwait.

Al-Asfoor, S. (1980) Composition and Origin of Residual Gravels in North Kuwait. M. Sc. Thesis, University of Kuwait, Safat, Kuwait.

Bagnold, R. A. (1933) A Further Journey Through the Libyan Desert. Geographical Journal, Vol. 82, p. 103–129.

Bagnold, R. A. (1941) The Physics of Blown Sand and Desert Dunes. Methuen, London.

Breed, C. S., Embabi, N. S., El-Etr, H.A. and Grolier, M. J. (1980) Wind Deposits in the Western Desert. Geographical Journal, Vol. 146, p. 88–90.

Bradshaw, M. J., Abbott, A. J. and Gelsthorpe, A. P. (1978) The Earth's Changing Surface. John Wiley and Sons, New York.

Butzer, K. W. and Hansen, C. L. (1968) Desert and River in Nubia. The University of Wisconsin Press, Madison, Wisconsin.

Chepil, W. S. and Woodruf, N. P. (1957) Sedimentary Characteristics of Duststorms: II Visibility and Dust Concentration. American Journal of Science, Vol. 255, No. 2, p. 104–144.

Cooke, R. V. and Warren, A. (1973) Geomorphology in Deserts: University of California Press, Berkeley, California.

Doehring, D. O. (1977) Geomorphology in Arid Regions. Donald O. Doehring, ed., Fort Collins, Colorado.

El-Baz, F. (1979) Siwa: Resort of Kings. Aramco World Magazine, Vol. 3, No. 4, p. 30–35.

El-Baz, F. (1984) Deserts and Arid Lands. Farouk El-Baz, ed., Martinus Nijhoff Publishers, The Hague, The Netherlands.

El-Baz, F. (1986) The Formation and Motion of Dunes and Sand Seas. In: Physics of Desertification. F. El-Baz and M.H.A. Hassan, eds., Martinus Nijhoff, Dordrecht, The Netherlands, p. 70–93.

El-Baz, F. (1988) Origin and Evolution of the Desert. Interdisciplinary Science Reviews, Vol. 13, No. 4, p. 331–347.

El-Baz, F. and Hassan, M. H. A. (1986) Physics of Desertification. Martinus Nijhoff Publishers, Dordrecht, The Netherlands.

El-Baz, F. and Maxwell, T. A. (1979) Eolian Streaks in Southwestern Egypt and Similar Features on Mars: Proc. Lunar Planet. Sci. Conf. 10th, Geochim. et Cosmochim. Acta Suppl. 11, Vol. 3, p. 3017–3030.

Foda, M., Khalaf, F., Gharib, I., Al-Hashash, M., and Al-Kadi, A. (1984) Assessment of Sand Encroachment and Erodibility Problems in Kuwait. Kuwait Institute for Scientific Research, Report No. KISR1297, Safat, Kuwait.

Gharib, I., Foda, M. A., Al-Hashash, M., and Marzouk, F. (1985) A Study of Control Measures of Mobile Sand Problems in Kuwait Air Bases. Kuwait Institute for Scientific Research, Report No. KISR 1696, Safat, Kuwait.

Gifford, A. W., Warner, D. M., and El-Baz, F. (1979) Orbital Observations of Sand Distribution in the Western Desert of Egypt. In: Apollo-Soyuz Test Project Summary Science Report, Volume II: Earth Observations and Photography, F. El-Baz and D.M. Warner, eds., NASA SP-412, p. 219–236.

Glennie, K. W. (1970) Desert Sedimentary Environments. Elsevier, New York.

Gore, R. (1980) Journey to China's Far West. National Geographic, Vol. 157, No. 3, p. 292–331.

Holden, C. (1991) Kuwait's Unjust Desserts: Damage to Its Desert. Science, Vol. 251, 8 March 1991, p. 1175.

Holm, D. A. (1960) Desert Geomorphology in the Arabian Peninsula. Science, Vol. 132, p. 1369–1379.

Khalaf, F., Kadib, A. L., Gharib, I. M., Al-Hashash, M. Z., Al-Saleh, S. A., Al-Kadi, A. S., Desouki, M., Al-Omran, L., Al-Ansari, L., Al-Houti, L., and Al-Mudhain, L. (1980) Dust Fallout (Toze) in Kuwait: Mineralogy, Granulometry and Distribution Pattern. Kuwait Institute for Scientific Research, KISR/PPI 108/EES-RF-8016, Safat, Kuwait.

Khalaf, F. I., Gharib, I. M. and Al-Kadi, A. S. (1982) Sources and Genesis of the Pleistocene Gravelly Deposits in Northern Kuwait. Sedimentary Geology, Vol. 31, p. 101–117.

Mabbutt, J.A. (1977) Desert Landforms. The Massachusetts Institute of Technology (MIT) Press, Cambridge, Massachusetts.

Mainguet, M. and Callot, Y. (1978) L'Erg de Fachi-Bilma, Chad-Niger. Centre Nat. Res. Sci., Paris.

Manent, L. S. and El-Baz, F. (1980) Effects of Topography on Dune Orientation in the Farafra Region, Western Desert of Egypt, and Implications to Mars. Reports of Planetary Geology Program, NASA TM-82385, p. 298–300.

Owen, R. M. and Nasr, S. N. (1958) Stratigraphy of the Kuwait–Basra Area: In Habitat of Oil, A Symposium, American Association Petroleum Geologists, p. 1252–1278.

Prospero, J. M. and Carson, T. N. (1972) Vertical and Areal Distribution of Saharan Dust Over the Western Equatorial North Atlantic Ocean. Journal Geophys. Res., Vol. 77, p. 5255–5265.

Safar, M. I. (1985) Dust and Duststorms in Kuwait. Meteorological Department of the Directorate General of Civil Aviation, Kuwait.

Sharp, R. P. (1963) Wind Ripples. Journal of Geology, Vol. 71, p. 617–636.

Skocek, V. and Saadallah, A. (1972) Grain Size Distribution, Carbonate Content and Heavy Minerals in Aeolian Sands, Southern Desert, Iraq. Sedimentary Geology, Vol. 8, p. 29–46.

Solle, G. (1966) Rezente und Fossile Wuste. Notizbl. Hess. Landesamtes Bodenforsch., Vol. 94, p. 54–121.

USGS (1963) Geologic Map of the Arabian Peninsula. U.S. Geological Survey and Arabian-American Oil Company (Aramco), compilers. U.S. Geological Survey Misc. Geol. Inv. Map I-270; scale 1: 2,000,000.

Weir, J. E. (1962) Large Ripple Marks Caused by Wind Near Coyote Lake (dry), California. Geological Society Am. Spec. Pap. 73, p. 1–72.

Whitney, M. I. (1978) The Role of Vorticity in Developing Lineation by Wind Erosion. Bull. Geol. Soc. America, Vol. 89, p. 1–18.

WIND EROSION AND ITS CONTROL

MONIQUE MAINGUET

Laboratoire De Géographie Physique Zonale,
University of Reims, France

INTRODUCTION

The first part of this paper will show how wind action becomes a damaging process in soil erosion and one of the most efficient mechanisms of land degradation. The second part will present actions and simple means to combat wind erosion and halt natural resource destruction, mainly short term curative measures for tackling a most severe environmental problem. Medium term measures, oriented towards the eradication of root causes of environmental degradation, are more difficult to apply because they concern socio-economical aspects of environmental problems.

It would be wrong to suggest that control of wind erosion can be accomplished by physical methods alone since there are considerable numbers of socio-economic and political factors, which will have a bearing on the success of control strategies and which must be considered before management techniques are implemented. For example, the consequences of conservation methods on the local people should be identified and the priorities of the land users evaluated. Any methods which are implemented must take into account several factors, including:

- the financial cost to local people of conservation methods and maintenance;
- the traditional organization of local communities;
- the time required for economic benefits to filter down to farmers;
- the access of farmers or nomads to traditional pastures or arable land;
- the distribution and marketing of agricultural products.

Wind erosion is particularly severe in dry ecosystems, arid and semi arid areas, where vegetation is insufficient to afford protection to unconsolidated surface material. A diachronic study of air photographies for 1953 and 1987 and enhanced SPOT images of 1987 along a transect stretching from the arid ecosystem south of Mauritania through the semi-arid ecosystem in Mali to the dry sub-humid ecosystem south of Mali and north of Guinea resulted in the following conclusions:

- Wind erosion is the exclusive damaging process in areas of up to 300 mm (annual precipitation) isohyet.

- Wind erosion and water erosion are the effective agents of degradation in areas from 300 to 750 mm of rainfall. Zhu Zhenda (1990) confirmed also the same thresholds in China (personal communication); wind can be harmful in these latitudes.
- Water erosion is the exclusive physical mechanism of degradation in dry sub-humid ecosystems when annual rainfall become higher than 750–780 mm.

Protection against wind erosion is based on control of shifting sand, active dunes or mobile dunes by the following three approaches:

1. Prevention of excessive removal of soil or sand particles by reducing winnowing of particles and their deflation.
2. Minimizing the transportation of sand by reducing saltation and creep. Appropriate techniques will be discussed for reduction in wind and speed near the ground.
3. Control of sand deposition in order to avoid human settlements by enhancement or deflection of the shifting sand, or promotion of the accumulation upwind of manmade installations.

SEVERE LAND DEGRADATION

The effects of wind are numerous. It is beneficial when used as a source of energy, for sailing, for wind mills, for pumping water, etc... It can also be a dangerous factor of soil erosion when it becomes a mechanism of particle deflation, of transport and a generator of dunes in areas previously used by man.

Wind and Aeolian Transportation of Sand

When pressure is the same throughout a system, the air remains in a state of equilibrium. Under normal circumstances, air, like any gas, behaves as a turbulent fluid and continuous variations in pressure occur owing to temperature differences and other variables or topographic features. Wind is therefore the air movement resulting from pressure differences between two points: the air flows from high pressures to low pressures, and the greater the pressure differences the stronger will be the resultant wind. On a global scale, horizontal air movements are more frequent than vertical ones.

Important progress has been made in recent years in the understanding of aeolian sediment transport. Much has been learned from two international workshops on the physics of blown sand held at Aarhus University, Denmark, in 1985 and 1990. Mathematical models have been established leading to a much deeper understanding of the physical processes involved in the initiation of motion of the particles and the transport rate. These help to better understand the geomorphological process which will lead to the highest efficiency in constructing protection against wind erosion by deflation, sand drift and sand encroachment.

The principal means of aeolian transport recognized are: **creeping, reptation, saltation** and **suspension**. Coarse sand moves by a process of creeping while medium and fine-textured sand is transported by reptation and saltation. Desert dust is carried by suspension, which is not dealt with in this paper.

Creeping

If particles are too large to be lifted by the wind, they can nevertheless move by rolling along the surface. Even pebbles and round blocks can be moved in this way. Creeping results from the direct effect of wind or from the impact of a grain falling against another. A grain falling at high speed can move another grain which is six times its diameter (more than 100 times its mass). The relative proportion of saltation and creeping is proportional to the diameter of the particles (Birot, 1981) and is shown in the following table. The percentage of grains transported by creeping in aeolian transport also varies.

Diameter of the grains and percentage transported by creeping and saltation (after Birot, 1980)		
Diameter in micrometer	**Creeping %**	**Saltation %**
150–250	16	84
250–830	25	75

Bagnold (1941) estimated that one quarter of the wind load is transported by creeping, while Chepil (1945) found that the process depended on grain size. He found that 15 percent of grains between 150 and 250 micrometers and 25 percent of grains between 83 to 250 micrometers are transported by creeping.

Reptation

In between saltation and creeping there is another mode of transport which is referred to as reptation (Anderson et al., 1990). Reptation is not saltation because the grains do not jump far from the surface nor is it creeping which corresponds to motion of grains without lifting.

Saltation

The maximum aeolian transport is effected by this process; saltation is linked with, and is responsible for, the other processes of aeolian transport, reptation and creep. In general particles smaller than 0.5 mm can be lifted obliquely or almost vertically and moved by the wind. The grains fall by their won weight along a curved trajectory at an oblique angle to the soil.

Saltation was defined by Anderson and others (1990) "as the transport mode of a grain capable of rebounding or of splashing up other grains.... When the wind blowing over a stationary sand bed becomes sufficiently strong, some, particularly exposed, grains are set in motion by the wind. Presumably, some grains are lifted fluid dynamically by the difference between the pressure below and above them, while the rest are pushed forward by the wind before they take off... Once lifted free of the bed, the grains are much more easily accelerated by the wind. Therefore, as they return to the bed, some of the grains will have gained enough forward momentum that on impact they rebound and/or eject other grains."

Saltation is a complicated process, the understanding of which involves a number of considerations:

The Value of the Fluid Threshold

According to Bagnold (1941 in Anderson and others 1990) the speed of the wind at which motion begins can be characterised by the means of the critical shear velocity u*

where: $$u*t = A\left(\frac{\phi s - \phi}{\phi} gD\right)^{1/2}$$

in which:
- ϕs = density of air
- ϕ = density of the grains
- g = gravitational acceleration
- D = the equivalent grain diameter
- A = an empirical coefficient depending on the function Reynolds number $u^*D\nu$
- ν = the kinematic viscosity of the air

According to laboratory studies the grains vibrate about a stationary position before being plucked up by the airstream directly into saltation. Unlike fluids, granular material requires a finite **shear force** to overcome Colomb (dry) friction between grains. For normal aeolian sand, an airspeed of 7 m/s is necessary to overcome the different **restraining forces.** The **mechanism of direct dislodgement** depends on grain size, shape, packing and sorting. **Threshold velocity** is the speed at which saltation occurs and varies with the particle diameter. On average the lowest threshold velocity occurs in particles with a diameter of 0.08–0.1 mm. Particles either larger or smaller than this require higher threshold velocities. Once initiated, saltating grains are capable of moving the same size of particles at lower wind speeds which can decrease until 3 m/s.

Factors Which Control the Amount of Sand In Transit

"In the initial period after the wind speed has exceeded the **critical value**, the sand transport is being built up and the number of saltating grains resulting from one impact is, on average, larger than one. Therefore, the number of saltating grains increases at an exponential rate. We call the average number of saltating grains, rebounding grains as well as *ejecta*, resulting from one impact, the **mean replacement capacity** ... As the transport rate increases, the time-averaged vertical wind profile is modified due to the considerable extraction of *momentum* from the wind by the grains in motion. The wind speed drops so that the saltating grains are accelerated less and impact at lower speed. As a result the mean replacement capacity decreases. When the mean replacement capacity has fallen to one, an equilibrium is reached. This equilibrium state is only stationary in a statistical sense. The number of saltating grains fluctuates around the equilibrium value... Computer simulations by Anderson and Haff (1988) indicate that the equilibrium is reached very quickly: in 1–2 seconds." (Anderson et al., 1990).

The height to which grains ascend depends on the initial air speed. Once the threshold velocity is reached, wind can transport more sand on a hard coherent surface than on a crumbly surface. In the same wind grains are lifted to different heights. The higher a grain is lifted, the more it is exposed to wind energy and the greater its impact on the soil.

A grain which is lifted vertically by moving air encounters air resistance in the opposite direction to the direction of air flow. During saltation, the grain is rotating and the angle of fall (α) is given by Bagnold's equation:

$$\frac{V^*}{V} = \tan \alpha$$

where: V^* = terminal velocity of the grain (m/s)
V = air speed (m/s)
tan = tangent

For a grain with a diameter of 0.3 mm, $V^* = 2.7$ m/s. In a wind of 12.7 m/s, the tangent is equal to 0.216 or 12°25′. For aeolian particles, the angle of fall is between 10 to 16°. As a result of these low values, the hollow and relief in the sand sheet will be accentuated as the grains fall on the upwind slope. This results in geant ripplemarks, as those of 1 meter high which can be seen in the Ubari sand sea (Libya), south of Sebah.

Initiation of Saltation

Saltation is initiated either by direct wind pressure beginning when the threshold velocity is reached or by the impact of saltating grains colliding with grains on the surface and then rebounding, a process attenuated by the fineness of the sand. The sand movement increases with the textural heterogeneity of the grains. After winnowing, saltation is increased because the grains of the top layer are coarser. The shape of the grains was shown to have an effective influence on collision (Willetts and Rice, 1986). Decreased sphericity of the grain results in lower velocities, shallower impact angles and fewer ejections by collision. Change of the angle of fall affects the effects of collision.

Local changes of bed slopes cause changes in the angle of impact: "With increased angle of incidence the ratio of **ricochet speed** to incident speed decreases and the ricochet angle increases... The vertical component of ricochet velocity increases with increasingly adverse bed slope. The forward momentum apparently lost at collision... increases with increased angle of incidence and this, it was suggested may be associated with increased creep activity." (Anderson et al., 1990).

The Height of the Saltation Cloud

At a windspeed of 5 m/s, saltating grains are usually in a band of about 10 cm thick, and occasionally as much as 1 metre thick, but this depends on the degree of cohesion and the roughness of the surface. The maximum density of saltating sand on a pebble surface is at a height of 2 m, but only 9–60 cm on a sandsheet. This has important implications for vegetation establishment and growth owing to the damage that can be caused to leaves and seedlings. For example, short grass will be unaffected by saltation if is growing on a pebbly surface, while young trees one to two metres high would be susceptible to damage. In contrast, the opposite effect would be seen on a smooth sand surface. During a sandstorm in the Ubari sand sea in Libya, saltation bounds reaching 4 m were observed on a substratum of coarse sand (Mainguet, unpublished results).

THE INFLUENCE OF TOPOGRAPHY ON WIND CONDITIONS AND WIND ACTION SYSTEMS

The Influence of Soil Parameters and Topography on Wind Conditions

Wind speed decreases near the ground as a result of surface roughness, whether caused by plants, natural or man-made obstacles, or just uneven ground. The resulting air turbulence consists of convergent, divergent and ascending air currents. Where the ground is vegetated, wind speed starts on average to decrease at ten times the height of the vegetation.

Soil surfaces influence conditions of sand movement. The topography and the texture of the soil surface affect saltation, which is minimized on a loose and fine textured soil and maximized on a coherent or coarse textured soil.

Middleton (1990) summarizes all the key variables in wind erosion system depending on the interaction between erosivity and surface erodibility. Wind erosion is reduced if the values of variables are increased (+) and if other variables are reduced (–).

EROSIVITY	
WIND VARIABLES	
Velocity	–
Frequency	–
Duration	–
Magnitude	–
Shear	–
Turbulence	–
ERODIBILITY	
DEBRIS VARIABLES	
Particle size	+ –
Soil clods and cohesive properties	+
Abradibility	–
Transportability	–
Organic matter	+
SURFACE VARIABLES	
Vegetation: residue	+
height	+
orientation	+
density	+
fineness	+
cover	+
Soil and moisture	+

Surface roughness	+
Surface length (distance from shelter)	–
Surface slope	+ –

Despite chemical fertilizers, soil productivity drops continuously in areas that experience regular dust storms (Fryrear, 1981). However, man does have the capability to use certain control measures against dust and sand erosion by the wind.

Bernoulli's application of Boyle's law (that the product of pressure and velocity is constant) of fluids gave rise to the Venturi principle. This states that **the mechanical energy of fluid motion is also constant. When applied to airflow, the result is that an increase in windspeed is associated with a corresponding decrease in air pressure.** Such a phenomenon occurs when streamlines converge, the air behaving as if it were flowing through a **nozzle**, with airspeed increasing and pressure falling. When streamlines diverge, as in a **diffuser**, the opposite occurs, i.e. a rise in air pressure and a fall in airspeed. **A hill or a shallow mountain as a dune profile have the same effect on wind (streamlines, pressure, velocity, sand transport) as a Venturi tube**. In fig. 1, the relation between streamline patterns, pressures, wind velocity and sand dynamics shows in:

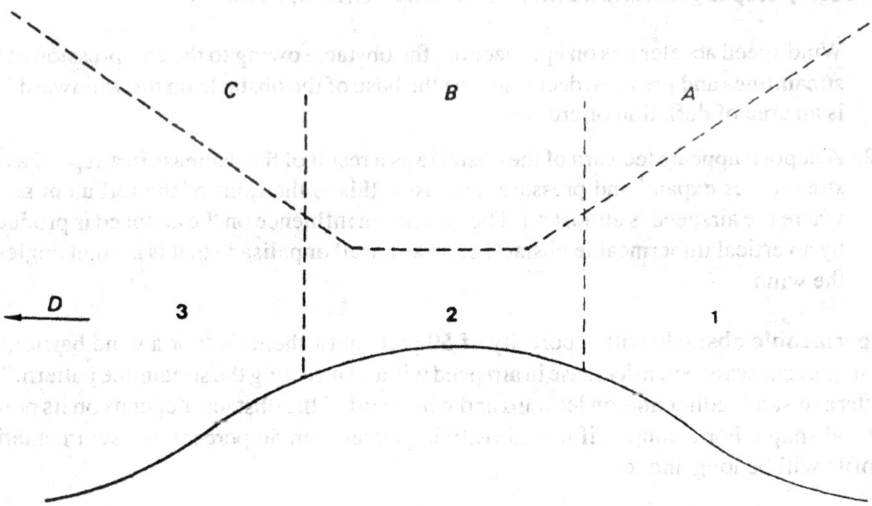

Figure 1. Venturi effect: model of wind speed and pressure changes over an obstacle (hill or dune). Wind speed is highest upwind and at the top and lowest downwind of the top of the dune.

A and 1. **nozzle** (upwind side)
- converging streamlines
- low pressure and high velocity
- area of erosion

B and 2. **throat**
- high speed
- transport area

C and 3. **diffuser** (leeward side)
- diverging streamlines
- high pressure and low velocity
- area of deposition

D
- Wind direction

1. In the **nozzle** (upwind side):
   – the streamlines are converging
   – the pressure is low and the velocity high
   – the area is an area of dominant erosion

2. In the **throat**:
   – the speed is high
   – the area is an area of particle transport

3. In the **diffuser** (leeward side):
   – the streamlines are diverging
   – the pressure is high and the velocity low
   – the area is an area of deposition

The model of air flow is by the Bernoulli rule: $P + \phi\frac{V2}{2}$

where: P = pressure
V = wind speed
φ = specific air mass

The effects of topography on air streams follows the same rules which depend on the profile of the obstacle and its permeability to air. For example, **an obstacle with gentle slopes** will give rise to the pattern of wind dynamics described in figure 2. In contrast **an obstacle with steep slopes** generates a different pattern of airflow, as follows:

1. Wind speed accelerates on approaching the obstacle owing to the compression of the streamlines and pressure decreases. At the base of the obstacle on the windward face is an area of deflation or erosion.

2. A deposit appears leeward of the obstacle as a result of the decrease in airspeed as the streamlines expand and pressure increases; this is the limit of the turbulent sector where the airspeed is almost nil. The maximum influence on the airspeed is produced by a vertical impermeable obstacle such as a wall or palisade that is at right angles to the wind.

A permeable obstacle with a porosity of 50 percent (a shelterbelt or a wind barrier, for example) causes an even decrease in airspeed without affecting the streamline pattern. The pattern of sand sedimentation leeward and windward of the obstacle depends on its porosity and shape. For example, if the porosity is greater than 50 percent, the sedimentation profile will be long and low.

Global Wind Action Systems

It is a widespread belief that wind in desert and semi-desert areas is dangerous because of its effect on the formation and migration of sand dunes. In contrast, the dangers of wind at the site of erosion and along the path of sand transport are frequently overlooked.

Combination of areas of sand deflation, sand transport and sand deposition detected on satellite images, thanks to their scale, allows the recognition of **Global Wind Action Systems (GWAS)**. **A GWAS**, as defined by Mainguet (1992), **is a dynamical aeolian system**

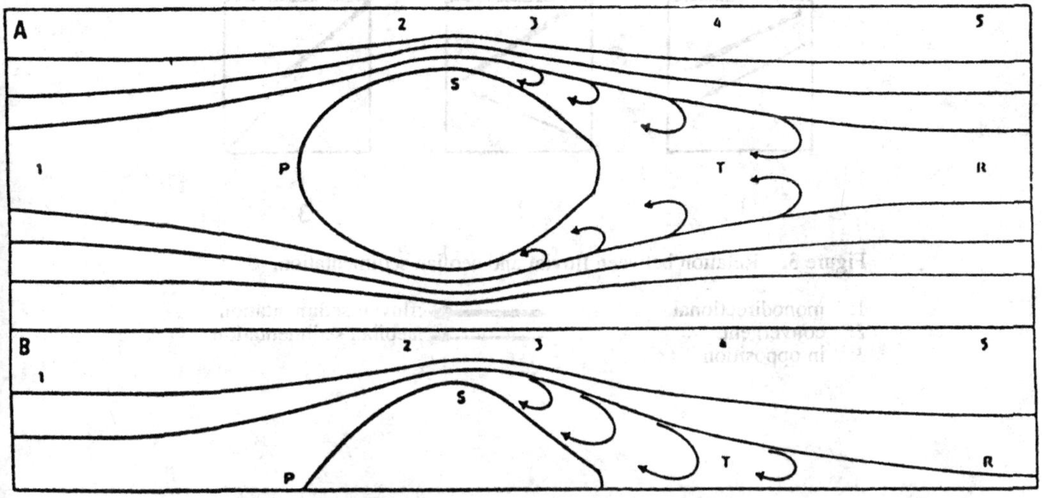

Figure 2. The effects of a gently sloping obstacle on the dynamic behaviour of wind streamlines and particles load; A, plan view; and B, cross-section:

1. Upwind section of the obstacle: the pressure is highest and the air speed is lowest, even nil at point **P**, at the bottom of the obstacle. Point **P** is an area of discharge and deposition.
2. Nozzle section: where the streamlines around the obstacle become dense on the sides. The airspeed increases and the pressure falls. This is a transport area, where wind erosion becomes very active.
3. Section of separation: the streamlines diverge at point **S**. The boundary layer separating from the surface is affected by vortexes, which produce a turbulent drag with higher wind speed. Deposition occurs at the outer limit of the drag area (point **T**).
4. The drag area behaves as if it were itself the obstacle. Its length and width do not depend on air but on the form of the obstacle. Leeward of this sector in **R**, the streamlines are not longer affected by the obstacle and they rejoin.
5. The section of rejoining is an area of sand deposition.

where in a definite area particles are exported, transported and accumulated or re-exported in consequence of wind activity. A GWAS can be an open or a closed system.

A **closed GWAS** is an area where particles are imported and accumulated but where export is negligible. The Taklamakan, because of its topographical configuration in a deep basin between high mountains, is probably the best example on the planet of such as closed GWAS.

A **open GWAS** means a system where, after import and accumulation, particles can be exported out of the system. The Sahara is the best example of an open GWAS; the correlations between Sahara and Sahel and how the Sahara exports its sand particles into the Sahel have been demonstrated by Mainguet et al. (1979, 1980a, b, Mainguet, 1991). The Tenere sand sea, for example, is prolonged without interruption by the Hausa sand sea in Niger which also receives a part of its sand from a branch of the wind current which runs west of the Air massif and from the winnowed alluvium accumulated by the *dallols*.

A **GWAS** can also be classified according to three models in which aeolian sedimentation is combined with a fluvial sedimentation or a marine source or particles. Figure 3 shows the relationship between fluvial and aeolian sedimentation.

Figure 3. Relation between fluvial and aeolian sedimentation.

1: monodirectional :fluvial sedimentation
2: convergent :aeolian sedimentation
3: in opposition

It should be stressed that sand dune formation is a phenomenon which can have three locations: along the sand currents, in the area of sand accumulation, or can be the end-product of wind erosion, and so, the final stage in land degradation. In the Sahel, the first stage is the loss of vegetation which exposes the soil to wind action, leading to loss of topsoil by the deflation of small particles usually less than 400 micrometers in diameter, while the residual coarse material may be reworked into active dunes. Dune stabilization must therefore be carried out concurrently with the conservation of the vegetation and the stabilization of the soil.

A difficulty is that wind action systems (WAS) exist at all levels: local, regional and synoptic (continental). A WAS can encompass either a very large or a small unit on the scale of a village. Local degradation can set in motion particles in areas previously fixed or vegetated and therefore stable. As a result of the deterioration of the natural vegetation cover by all types of overuse (grazing, firewood cutting, shifting cultivation, excessive trampling along livestock trails, vehicles tracks, etc.), the surface will be exposed to wind action and fine particles will be exported by wind since the coarser ones will be reworked and form ripples or dunes. The sand movement in this case is relatively local with an erosion area of a few square kilometers. However, if human settlements are located on a regional or synoptic WAS (a GWAS), the local WAS induced by human activities will be included in the natural global WAS. Around the oases south of the Taklamakan desert, Hotian for example, a local WAS have been created. The situation has been aggravated in recent years by population growth and because this localised WAS is located on the more global WAS—the GWAS—of the whole Taklamakan desert. In the Eastern Province of Saudi Arabia, within the Jafurah sand sea located in an arid ecosystem (less than 80 mm of rainfall per year) and where the sand drift according to Fryberger and others (1984) reaches 30 m^3/m width in relation with the oilfield and petroleum installations have developed industrial complexes, residential settlements, transport and communication infrastructures. All these achievements disrupt the sand flow and stimulate sand accumulations.

Another difficulty is to define the limits of a WAS. It is not limited—but is nevertheless influenced—by topographical constraints: small or medium obstacles do play a role in wind processes but are not real inhibitors of wind action.

Regional or Synoptic Wind Action Systems

Source Area

Wind action systems at a regional or continental scale are those where the erosion site may be hundreds of kilometers from the area of deposition. In many deserts the scale of the aeolian systems caused by moving sand is so large that it precludes only long-term solutions. The source of sand is distant, the size of the sand flow so large that stabilizing the system cannot be achieved, but the system must be understood to properly deal with it.

The source area can comprise inexhaustible coastal sand. Two good examples are given by Essaouira, in central Morocco and Layyoune, in south Morocco. In both cases, the marine sand of the Atlantic coast feeds encroaching sands currents which disturb the development of these areas. The same case can be observed along the Indian Ocean coast of Somalia, where, south of Mogadisho, the little towns of Brava and Shalambot are constantly submitted to invasion of aeolian sand, the source of which is the old red coastal dune, itself fed by marine sand.

The source is an area where loose particles are deflated by wind of sufficient speed. The first manifestation of wind erosion in such areas is the dust storm which consists of small particles of clay and silt with diameter of between 2 and 50 micrometers. All continents experience them, the more powerful of which can have plumes 500–600 km wide and 2,500 km long (Péwé, 1981). All mineral aerosol size distributions indicate a maximum between 0.06 and 0.08 micrometers in size (D'Almeida and Schultz, 1983).

Dust can be carried thousands of kilometers by the wind. For example, dust is carried from the Sahara towards northern and central Europe (Bücher, 1986), the Atlantic Ocean, the Caribbean or even towards Kazakstan in the ex-U.S.S.R. Mineral dust is transported from arid regions of North Africa into the tropical north Atlantic during the summer months. The dust often reaches 5–7 km altitude, spreads over hundreds of kilometers in latitude, and extends to the Caribbean Sea and the southeast United States. In winter, large quantities of soil dust are transported primarily to South America (Propsero et al., 1981). The annual mass of Saharan dust transported over the North Atlantic Ocean is estimated to be about 260 million tons.

The severe drought currently afflicting the Sudano-Sahelian zone to the south of the Sahara Desert has been suggested as being instrumental in producing an increased input of soil-derived aerosols into the atmosphere from the region. During the very dry periods 1972–74 and 1976–1984 mean aerosol concentrations at Barbados, West Indies, as affected by the African dust plume were three times that of pre-drought levels, that is before 1968. A marked increase in the frequency of severe dust occurrences in northern Nigeria has also been noted during 1972 and 1973. Data from selected meteorological stations which show that dust-storm activity in the east and west of the Sudano-Sahelian belt has dramatically increased during the drought years are presented by a factor of six in Mauritania and up to a factor of five in Sudan (Middleton, 1985).

The clearing of land for agriculture is one cause for the loss of topsoil by winds deflation. All areas, seasonally or permanently, vegetation-free or nearly vegetation-free, in arid and semi-arid regions, are potential dust, silt and sand source areas. In semi-arid regions, all devegetated dunes and sand sheets are also vulnerable. An example is the overgrazed vegetated *red dune system of the Sahel* which stretches from Senegal to Mali, Niger, Chad and Sudan, and has become as a result of desertification, a new source area. In the past, the Sahara was a part of the sand source for the Sahel which was the area of deposition. With increasing desertification, the Sahelian sand sheets are now being destablized, and reactivated and the Sahel is now also becoming a source area. Long-term programmes of immense dimensions would be required to yield significant sustainable results in protecting areas from dune encroachment.

Areas of Sand Transportation

Sand movements are detectable in the field, in aerial photographs and satellite images because, when transported, the sand is organised in **sand currents** at the interface of the soil and the atmosphere. Along these sand currents the movement of sand particles can be:

- a **collective movement organised in dunes:** *linear dunes* along which sand moves as along railways; or *crescent dunes* (barchanic edifices or barchans). These dunes are the only one which migrate as one unit with all the grains participating in the movement. Both crescent dunes and linear dunes can be classified in the **family of sand transport edifices.**
- a **movement of a population of individual grains**, by *reptation, creeping* and *saltation*, which is detected because of collisions of the grains on the surface. This impacting has an *abrasive effect* called *corasion* and leaves on the rocks characteristic lines and grooves which are in arid zones more reflectant than the general desert patina which darkens the rocks.

Watson (Goudie, 1990), summarizes the approaches which have been employed to reduce drifting sand:

- **Promotion of the deposition of the drifting sand** by
  - **ditches** sufficiently deep to prevent aeolian scouring of sand from the floor and regularly cleared of the sand which accumulates;
  - **barriers** and fences which create sand deposition and therefore reduce the sediment load of the airstream and sand mobilization downstream because saltation is reduced upstream.
- **Plantation of vegetation belts to trap the sand moving away from source areas.** Bushy shaped and fast growing plants are recommended and irrigated plantations in arid and semi-arid ecosystems.
- **Enhancement of the transportation of sand** by
  - **shaping the land surface** to increase the wind velocity over the surface and the removal of particles. According to the Bernoulli rule and the Venturi model the airflow on a windward slope is compressed, its velocity increases and the wind capacity to move the sand up the slope also increases.
  - **treatment of the surface**. A smooth resilient surface does not absorb the kinetic energy of the saltating grains (Watson, 1990).
- **Deflection of the sandstream** with fences planted about 45° to the direction of sand drift. Each face has an angle of 30° to the wind direction.

Areas of Sand Accumulation

Dust is exported long distances while sand is carried shorter distances and can be reorganised in dunes within the sand flow or accumulated as dunes within the **deposition area**. Any cause of wind reduction results in deposit of the load of sand. This includes:

- topographic obstacles;
- new surface roughness introduced by a natural or man-made vegetation cover; and
- surface roughness introduced by a sandy cover.

After depositing its load, the wind increases its energy which allows it to take on a new load but one of finer sand. This phenomenon could be called **load substitution**. Controlling sand movement is often considered as **dune stabilization** in the narrow sense but is also controlling the movement of particles from the source area to the transport and accumulation area.

GENERAL SOLUTIONS

In the erosion area, a knowledge of the location, nature and extent of both the erosion and the sand movement area is necessary. The source area can be very large indeed as for example in the Egyptian desert where the wind current bringing the encroaching sand to the Kharga Oasis originates in the Abu Muharik dune system, which is over 300 kilometers to the north-north-west.

In many sites, it may be cheaper and more effective for a project to take measures that will prevent the sand from being picked up in the first place than to fix the dunes formed in the accumulation area. In the erosion areas, the objective is to protect the soil particles from deflation and this can be achieved by:

- reducing wind speed by the establishment of windbreaks of dry or preferably living vegetation if water or rainfall permit; the strategies are different according to the degree of aridity;
- increasing the cohesion of soil particles by planting vegetation; and
- fixing the soil using chemical adhesives, or by covering the loose particles with plastic sheets, nets or dead vegetation.

In the **transportion area**, the objective is to bring about a reduction in wind speed by increasing the roughness of the land: the resulting turbulence causes the airborne particles to settle. Deposition of sand in pre-selected areas will reduce the threat to human settlements.

In the **deposition area**, sand sheets or dunes can form. The different shapes of dunes are related to sand supply, windspeed, the duration of the wind, the wind regime, and the local topography. Several types of sand dunes exist and the dynamics and control of the most important ones will be discussed.

CRESCENT AND LINEAR DUNES

Dune Classification

In an area where the importation of sand surpasses its exportation, active sand edifices are formed called **depositional dunes**. Their volume—but not their height—and number increase as long as the feeding of sand is in excess. When in an area of abundant sand, exportation becomes the dominant process, other sand edifices appear, for which the proposed name is **erosional dunes**. This name is selected because of the characteristic form and direction, which result from deflation, i.e., from erosion and aeolian exportation.

Degrees of activity and classification of dunes

Dunes can also be classified into three categories according to decreasing activity: **active dunes, fixed dunes** and **vegetated dunes**. The term **active dunes** is reserved for constructions of aeolian material independent of the size of the particles (there are dunes of clay, of silt, of sand, or of mixtures of two or three of these), or of the nature of the grains.

- A sand edifice loses its characteristic activity and becomes a **fixed dune** as soon as pedogenetic mechanisms intervene or when the dune particles are aggregated by iron oxides or silicium, or encrusted by gypsum or carbonates. A mechanical fixation results from winnowing with the formation of a pavement of coarse sand particles with dimensions exceeding those exportable by the wind.
- The third category consists of the **stabilized** or the **vegetated dunes**. When climatic changes occur, or after human-made degradation, the process of reactivation occurs in such dunes.

In continental areas, the limit between active and vegetated dunes correspond to the 100 mm isohyet when the rainy season is also the cold season and the 150 mm isohyet when the rainy season is the hot season. Active dunes do not occur naturally when rainfall is above these limits. They result from land degradation, when the vegetation cover is depleted because of a drought or human over-use. Then the sand is loosened and the previously stabilized dunes becomes a sand reservoir that feeds active sand and can be dangerous to the human settlements leeward of the area.

Wind Regime and Dune Classification

In 1931, Aufrère, in his *Morphological cycle of dunes*, proposed a dynamic classification of dunes in which he distinguished among:

- *Winds in conjunction*: This wind regime occurs when the dominant direction diverge, scarcely distant from one another. This is the regime of *trade winds* that Fryberger (1979) called **narrow and wide unimodal**. The distinction between a narrow and a wide unimodal wind regime is in the oscillation of winds forming a small angle from one specific direction. It can also include many wind directions where only one is effective.
- *Winds in opposition*: This wind regime occurs where the winds have two dominant directions at an angle of about 180°. Examples of this are the *Chergui* and the *Saheli*

of meridional Morocco, the *Alize* and the *Sirocco* in south Algeria and Tunisia as well as the monsoon and trade winds of the Sahara.

- *Incident winds*: These are bidirectional regimes that are diverse and are sub-divided into numerous branches by topographical irregularities and thus cannot be discerned by a network of anemometers. This is the **acute and obtuse bimodal** regime described by Fryberger (1979).
- *Multidirectional winds*: These are the wind regimes with at least two dominant winds that create a **complex** regime.

Thus, the active dunes can be classified into three families corresponding to three principal aeolian regimes: (a) **crescent dunes** and **transverse chains**, response to an anemometrical regime that is dominated by a monodirectional wind; (b) **linear dunes**, the result of an anemometrical wind regime that is continually disrupted by topographic irregularities or bidirectional winds, and (c) **star dunes**, the creation of a complex anemometrical regime without a dominant wind direction.

Sediment Balance and Classification of Dunes

It is not enough to classify the dunes only according to their degree of fixation, the wind regime and their dynamics. It is necessary to consider two other fundamental parameters, the sediment balance and the directions of accumulation and exportation. Thus, it is here proposed to distinguish between: (a) **active depositional dunes** including barchanic edifices, all transverse dunes, linear dunes, and star dunes, corresponding to a positive sediment balance (**P.S.B.**); and (b) **erosional dunes** which are deflation forms, including parabolic edifices, and sandridges corresponding to a negative sediment balance (**N.S.B.**) as discussed by Mainguet (1984).

When a vegetated sand sheet with sandy hills is locally degraded, blowouts, which are deflation forms, appear on the windward slope of the hills and the sand is deposited on the leeward slope and parabolic dunes develop. Parabolic dunes in their first step are erosional dunes corresponding to a "Negative Sediment Balance" (N.S.B.) In the Radjastan desert a sufficient volume of sand is collected so as typical U-shape and V-shape composite parabolic dunes are formed with several tens kilometre long arms. These dunes evolve by elongation of their arms and slow migration of their body.

Organisation of Transverse Dunes

Aeolian features can be classified as isolated edifices or groupings of dunes, in which one is able to distinguish their organization and forms that are separated by open ground. In each subclass of dunes distinctions can be made between texture or form, and structure or pattern. In the classification of Breed and Grow (1979) a distinction was made between (a) **simple** (or **elementary**) **dunes**, the most fundamental forms, (b) **composite dunes**, which comprise the same fundamental elements of the simple type but they occur in greater numbers or in different sizes. The composite forms are associations of individual components which can be of equal or unequal dimensions; and (c) **complex dunes**, which are composed of many varieties of different dunes, combined by juxtaposition or superposition.

The principle dune patterns are: (a) **isolated** dunes where the placement of each structure is independent of its neighbors; (b) the **alignment** of dunes based upon a preferential direction; and (c) **groupings** of dunes where alignment is in several directions, in triangular and checkerboard patterns.

All the damages caused by wind can be classified into three main categories:

- Wind erosion which can also be called **wind ablation, deflation,** or **winnowing** and **corasion** which all mean an abrasive action.
- Sand accumulation.
- Transport of particles and the genesis of a family of dunes which themselves participate in aeolian transport of sand particles, are crescent dunes called barchans and linear dunes called seif. The latter two are discussed below:

Barchans

Barchans represent a family of mobile dunes (Fig. 4). In their initial stages, which we call **pre-barchanic,** they are stationary and therefore do not constitute an immediate threat whilst others, including the true barchan and the **barchanic dihedron,** are dangerous, essentially because they do migrate.

Geomorphology of the Barchans

Barchans are dihedral crescent-shaped dunes with a convex windward face and two oblique horns, or wings, which tail leeward and are extended in strong winds. A perfectly symmetrical formed barchan is rare, as it is usual to find horns of different lengths. The dunes can vary in length from 50 centimeters to more than one kilometer and are between 0.3 and 50 meters high. The two faces of a barchan, rejoining in a sharp crest (brinkline), are topographically and dynamically different: the windward convex slope is a face both of deposit and removal and it has an angle varying from 6 to 12 degrees. The leeward side of the dune (the slip face) is much steeper, between 22 and 33 degrees, and it is a slope mainly of deposition and slippage.

The conditions for barchan formation are: a wind regime with one prevailing wind; a flat topography; availability of sand with a diameter between approximately 0.125 and 0.35 mm; a cohesive substratum of coarse sand, gravel or rock; a dry surface without vegetation or a very sparce steppe.

Barchans or barchanic edifices can exist in isolation or in lines adjoining crescents known as **barchanic chains**. These may be clearly defined or rather erased when they are known as **transverse chains** or **transverse dunes**. The latter are characteristic of sandy deserts, with a rich sand supply, the most prominent example being the Taklamakan desert in Xinjiang (China). Barchans are infrequent in the Sahara, where the sand supply is not abundant, except on the semi-arid northern and southern boundaries.

Barchan Genesis and the Formation of Pene-barchanic Structures

Prior to the formation of a true barchan, successive stages are common. **Dome dunes** can form without topographic obstacle as a result of the horizontal or vertical diffluence of sand-bearing winds. A prevailing wind over a **sand dome** results in an area of low pressure

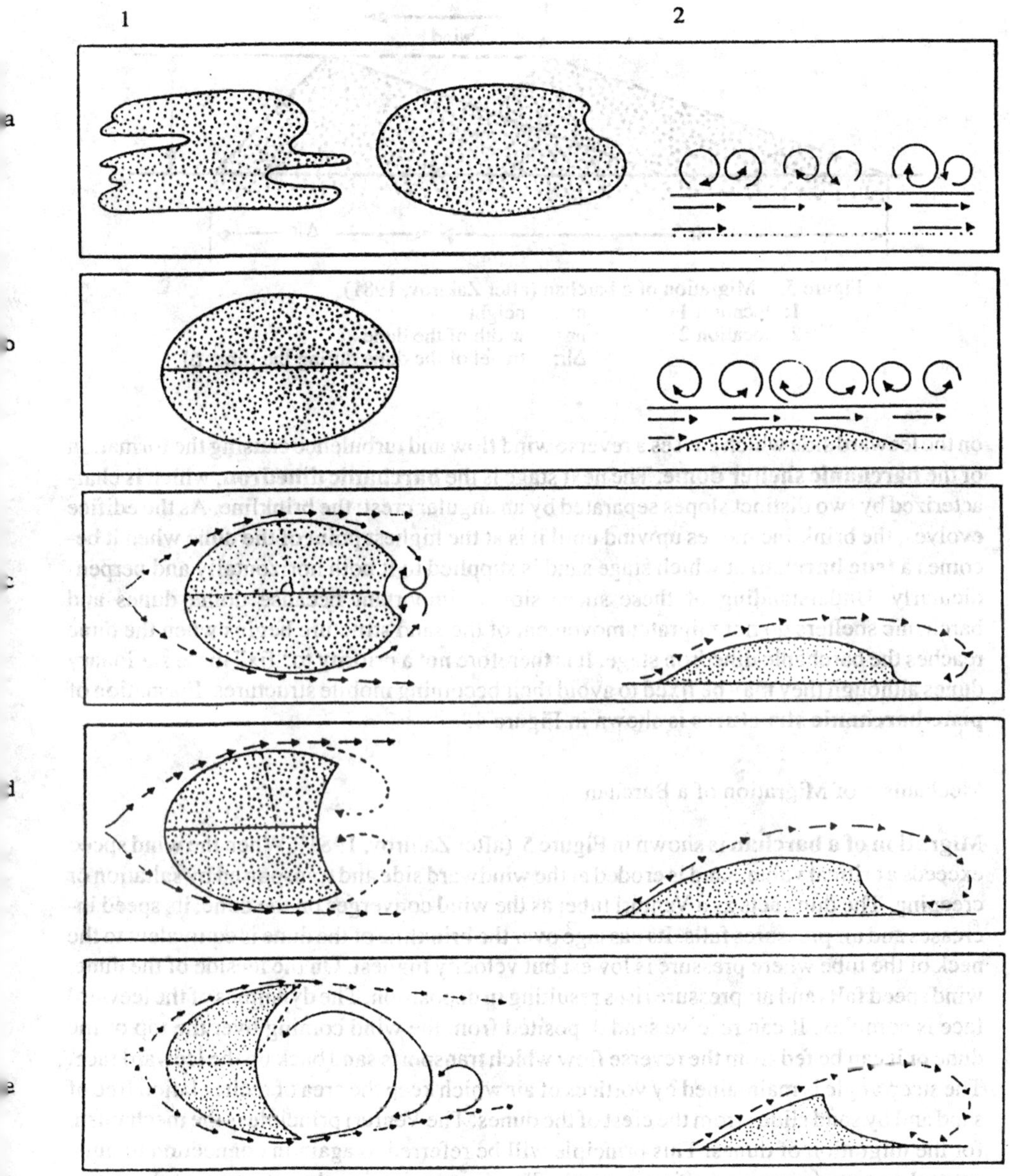

a : irregular aeolian sand veneer
b : aeolian sand dome
c : barchanic shelter
d : barchanic dihedron
e : barchan

1 : plain view
2 : cross section
: turbulent layer
: laminate layer
: dominant wind

Figure 4. Pene-barchanic structures.

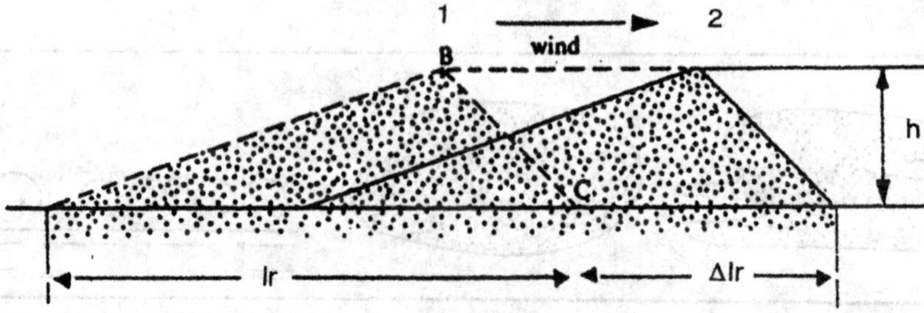

Figure 5. Migration of a barchan (after Zakirov, 1981)
1: location 1
2: location 2
h: height
lr: width of the dune
Δlr: travel of the dune during the time Δl

on the leeward side which causes a reverse wind flow and turbulence causing the formation of the **barchanic shelter dome**. The next stage is the **barchanic dihedron,** which is characterized by two distinct slopes separated by an angular crest: the brinkline. As the edifice evolves, the brinkline moves upwind until it is at the highest point of the dune when it becomes a **true barchan** at which stage sand is supplied to it both horizontally and perpendicularly. Understanding of these succession is important because dome dunes and barchanic shelters do not migrate; movement of the sand structure begins when the dune reaches the barchanic dihedron stage. It is therefore not a priority to "fix" these stationary dunes although they may be fixed to avoid their becoming mobile structures. Formation of **pene-barchanic structures** is shown in Figure 4.

Mechanism of Migration of a Barchan

Migration of a barchan is shown in Figure 5 (after Zahirov, 1981). When the wind speed exceeds a critical value, sand is eroded at the windward side and transported by saltation or creeping. The dune acts as a Venturi tube: as the wind converges on the dune, its speed increases and air pressures falls. Its passage over the brinkline of the dune is equivalent to the neck of the tube where pressure is lowest but velocity highest. On the leeside of the dune, wind speed falls and air pressure rises resulting in deposition. The dynamics of the leeward face is complex. It can receive sand deposited from the wind coming over the top of the dune or it can be fed from the reverse flow which transports sand back up the leeward face. The steep angle is maintained by vortices of air which keep the area of recirculation free of sand and by sand slides from the crest of the dunes. The Venturi principle is the mechanism for the migration of dunes. This principle will be referred to again in connection to dune control and sand sweeping effect on roads. The dune maintains almost constant shape and direction, although as its size increases, the speed of migration decreases.

The speed of dune movement is proportional to sand transport rate, i.e., to cubic wind velocity when this velocity is higher than the threshold velocity and in inverse proportion to the height of the edifice. Dunes which are wet, do not move until the sand surface is dry.

With similar wind speeds and sizes, individual crescent dunes move more rapidly than dunes which are grouped in transverse chains. That explains why on the border of the sand seas, where the oases are often located, the dune movement is faster than in the center of the desert. This trend is confirmed around and in the Taklamakan and Badain Jaran deserts: in the desert the dunes reach a speed lower than 5 m/year and more than 10 m for the scattered dunes in Kashi Delta oasis, in the southwest of Pishan oasis, in southwest of Qiemo in Taklamakan desert and in Anxi and Minqin oases in Hexi Corridor (Zhu, 1990).

In a flat area with a prevailing wind which is conducive to barchan formation, the dune will migrate unimpeded. It can also climb slopes but in so doing the dune changes its form and may become a linear dune. After the obstacle has been passed, the barchan reappears (Mainguet, 1972). This permanent barchanic trend is an important one and an example may be seen north and south of El Karga oasis in central Egypt. There the barchans cross the desert and on arrival at an oasis, village or plantation, they break down. Once away downwind from the oasis, the sand flow reforms into barchans (El Baz, 1984).

A barchan loses its form when it meets an obstacle; its material is blocked and becomes a hindrance. This occurs when a barchan reaches a vegetated area such as a palm tree plantation or an oasis, or a natural obstacle such as a steep hill escarpment. Similarly, when the dune reaches any obstacle which divides the wind stream, it will disappear and a linear dune can appear leeward of the obstacle until the barchan spontaneously reforms downstream. A group of topographic obstacles can result in the total disintegration of a barchan but not of the flow of sand.

Methods for Preventing Barchan Migration

Four methods of halting barchan migration are feasible: fixing the sand in the source areas; creating of an artificial dune; fixing the barchan itself; and destruction of the barchan.

1. Fixing the Sand in the Source Areas

A source can be fixed if it has a small surface area but this may be difficult or impossible where the source of sand is large or in a region far upwind of the dunes, perhaps in a drier ecozone. The source may also be the surface along the path of the barchan. For example, the Kharga oasis is threatened by the entire desert upwind of Kharga because the substratum itself has produced loose sand by corasion or deflation. This is common where there is a basement of sandstone or fluvial deposits.

2. Creating of an Artificial Dune

To avoid the arrival of huge amounts of sand, which endanger human settlements, the creation of an artificial dune can be the solution. The objective is to encourage sand accumulation windward of the human settlement. This method requires the building of a palisade or wall at right angles to the main wind direction 150 to 200 m upwind of the area to be protected. The first palisade, which forces the sand to accumulate, will, if it is efficient, be readily buried by sand deposition, therefore it must be lifted and another palisade parallel to the first one should be built 100 m from the settlement. This technique is used with success in Hannabou (Morocco) where a biological belt was planted parallel to the palisade.

The creation of an artificial dune occurs when the arriving barchan is broken down by the use of temporary physical barriers erected in parallel, or by placing permanent palisades in a lattice-work. It must again be emphasized that a single barrier constructed downwind perpendicular to the axis of symmetry of the barchan is useless, as it will be buried by the advancing dune. The resulting artificial dune grows and as it does so, the fences must be raised. This stage must be followed by biological dune fixation as explained above.

North of Essaouira (Morocco), along the coast, an artificial dune has been created by establishing little fences parallelle to the coastline. The same method is used in Somalia to protect the town of Brava. In both cases the wind is oblique to the coast. The initial strategy based on fences parallele to the coastline had to be modified. To play a better role of sand trapping, the fences had to be reoriented at right angles to the oblique coastal wind.

For example, Golinski and Lindbeck (1979) have used dune forming fences in woven plastic along the southern coast of north south Wales, Australia, to accumulate sand into a dune form preparatory to establishing vegetation. Woven plastic is cheap, quick and easy to erect, and has therefore replaced brush, slat and jute mesh materials previously used.

3. Fixing of the Barchan Itself by Immobilization

All the methods used in fixing a dune are based on the stabilization of the surface of the dune: a complete mantling, stabilized strips (the strips are at right angles to the axis of the dune) or fixing the windward face of the dune to prevent sand being carried from these to the leeward face. The technique used depends on the biological environment and the availability of *inexpensive local products* such as palm fronds or other dry vegetation. Plastic nets and chemical products are generally expensive and should be used as the last solution.

In addition, palisades must be placed upwind of the dune to trap incoming sand and to cut off the sand supply to the dune. Because the vertical growth of a sand accumulation along each side of a palisade is not definite when such an accumulation has reached its equilibrium height, the sand no longer trapped, is transported further ahead and a second palisade must be erected.

The most convincing explanation of why the vertical growth of a dune or a sand accumulation is not indefinite is in the work of Lai and Wu (1978 in Watson, 1990): accumulation at the crest increases the height of the dune and therefore steepens the windward slope increasing the wind speed. As a result, Watson (1990) shows that the sand cannot accumulate indefinitely because the leeward slope becomes oversteepened and unstable and avalanching occurs when the slope reaches the angle of initial yield which is between 31° and 34° for the aeolian sand. Simultaneously at the steeper higher portion of the deposition form the wind velocities are so high that the rates of sand transport are greater and the particles are transported over the dune. At this step a second palisade parallele and leewards to the first becomes necessary.

Dune stabilization using revegetation is the best method because it ensures effective long-term protection against the wind, with plants acting as a natural trap for airborne sand. However, plant introduction requires care and plants adapted to local ecological conditions should be established in nurseries. Failures have occurred by sowing tree seeds directly on dunes: germination takes place while the sand is moist but the abrasive effect of moving sand kills the young seedlings. The sand should be immobilized mechanically long enough for the establishment of vegetation, but once established, plants will improve soil conditions by lowering soil temperature, anchoring the soil and improving soil structure and chemical conditions. However, soil moisture will fall owing to evapotranspiration although this will be offset to some extent by the increase in soil organic matter which tends to have a moisture retention capacity. Fertilizers and irrigation may be necessary and it is important to exclude livestock during the whole establishment period.

4. Reshaping or Destroying of the Barchan

Reshaping of the dune means first to alter its aerodynamic barchanic form by cutting the horns which supply sand to the dunes downwind.

The traditional and least expensive method is to scatter large piles of rocks on the windward side and on the brinkline of the dune. This technique has been used in the area of Merzuga (Morocco), and oasis of the Wadi Draa. An alternative method is to change profile of the dune by causing a dispersing of the sand.

A barchan can be removed by mechanical excavation and transportation of the sand. This expensive method is realistic only if the sand can be used in manufacturing of building material or road surfacing material, for example, Liu Shu (1986) describes in China how barchans have been dissipated thanks to grass seedlings in the interdunal areas.

Combatting Migrating Barchans In Semi-arid Ecosystems

The technique of sand fixation differ according to the local and to the type of edifice. To fix reactivated sand or active sand in semi arid and in dry sub humid ecosystems, the Chinese use systematically straw checkerboard or cover the sand with clay or gravel before implementing the second step of biological rehabilitation. To stop migrating barchans the methods used in the Chinese dry areas can be taken as a model for all other regions threathened by barchans. The Chinese method consists of three steps:

- **first step**: trees are planted in the interdunal surfaces. These trees have a blocking effect in relation to the migrating dunes, whereas on the convex back of the dunes are planted bushes (*Artemisia*, for example)
- **second step**: when the sand is stabilized, willows are planted on lower 2/3 of the dune. In the meantime the wind blows away the sand of the upper part of the dune which becomes a convex edifice. Simultaneously as the bushes planted on the back of the dune grow, the dune loses its capacity of migration and becomes a motionless low edifice.
- **third step**: the dune can now be reafforested in its totality.

Seif or Linear Dunes

Seifs, or linear dunes, are another type of active dunes even more difficult to understand and to stabilize than barchans. While barchans move as a whole, linear dunes are dangerous because as they grow, they elongate as much as 100 meters per year, cutting roads and railways or covering houses, gardens, wells and agricultural land.

Geomorphology of Linear Dunes

Linear dunes are narrow, sinuous in the detail, but run in a straight line in general, and structures which vary in length from some tens of metres to many kilometres. In the *Grand Erg de Bilma*, in Niger, a large seif reaches 40 kilometers long, 50–100 meters wide and 30 meters high. In the initial stage they are a straight **sand arrow** but as they grow, they appear as a line of open crescents linked in sinusoidal manner. In the deserts of Western China, this sinuosity is so pronounced that the dunes are known as **fishing hooks** or **honeycomb** linear dunes.

These dunes are usually asymmetrically triangular in transverse section with two slopes separated by an angular crest. The windward flank is straight or slightly convex, while the leeward flank can be slightly concave and has a steeper slope. In some areas, such as in the deserts of Egypt, the asymmetry of the linear dunes reverses seasonally, probably because of the seasonal changes of one of the two prevailing winds. Field observations and meterological data show that linear dunes can occur in areas where there are two seasonal prevailing winds coming from different directions.

Seif dunes exist in three forms:

- **Symmetrical seifs** or **sand arrows**. These are sharp-pointed on the leeward end and have two faces on either side of the crest-line along which they are bilaterally symmetrical.
- **Asymmetrical seifs**. These have a more developed windward slope and a slip face leeward and they tend to be sinuous.
- **Seif dunes with alternating slip faces**. These occur where there are seasonally changing winds.

Location and Formation of Linear Dunes

The siting of seifs is a complex aspect of these dune types. They can be anchored on the leeside of an obstacle such as a hill or an escarpment. A good example of this is to be found in the oasis of Kaouar in Niger: the linear dunes are formed at the foot of a sandstone escarpment, thereby limiting its extent upwind. An interesting example can also be seen in the Navajo Indian Reservation in northern Arizona where linear dunes of 2 to 5 km long run towards the north east from the escarpment which separates Ward Terrace and Moenkopi Plateau.

Alternatively, seifs can appear on coarse sandsheets without a topographic obstacle. The sand is the residue after winnowing of the fine particles of the sand sheet. They may also be juxtaposed or superimposed on other dunes. In the Simpson Desert (Australia), the seifs meander along the top of a substratum of sandridges and snow characteristic junctions which are open upwind.

Bagnold (1941) suggested that seif dunes could be formed by the asymmetrical extension of a barchan. When an aeolian wind flow meets an obstacle, an eddy forms in the obstacle's lee and sand is deposited in dunes at a tangent to the eddy area. Linear dunes, that do not originate from barchans or obstacles, the formation of which has yet not been explained, can also be observed: south of Sebah (Libya) they are the general states.

A linear dune evolves by elongation and means of elongation must be understood in order to control such dunes. The wind crosses the windward flank obliquely while on the leeward flank the wind circulation is complex causing sand deposition and "avalanching" which explains the steepness and the loosely packed sand on this flank. In addition, a reverse flow occurs on this leeward flank, generating a spiral action which smooths the sand and transports it both obliquely and longitudinally along the dune causing ripple marks at right angles to the crest line. Tsoar (1985) observed that the flow is divided into two components, one of which is parallel and the other perpendicular to the crest.

The evolution of linear dunes therefore takes place by three simultaneous processes:

- Sand reaches the dune where some of it is deposited on the windward face:
- More sand crosses the crest to the leeward face where it is caught in the eddy and brought back to the dune and deposited.
- Sand is transported along the flank and the foot of the dune and migrates parallel to the dune axis causing progressive elongation.

Control of Seif Dunes

Control of linear dunes is difficult because they elongate at an oblique angle to the prevailing winds. Three methods are feasible:

1. Preventing the sand from arriving at the dunes

This is achieved by reafforestation of the area upwind of the dunes. If the source area is too large and if reafforestation or revegetation is not possible owing to lack of water, an alternative is to deflect the wind by placing shelterbelts at right angles to wind direction, **NOT** at right angles to the dune.

2. Halting the longitudinal migration of sand using barriers

Barriers are placed in a **herring-bone** pattern at right angles to the prevailing wind in the corridors between two dunes (Fig. 6). These intercept and deflect the sand. They can be constructed from local materials such as millet or grass stalks, twigs, brushwood, or palm fronds but shelterbelts made from living vegetation are better. The distance between two barriers should be ten times the height of the fence. This method is being used in Shalambot, south of Mogadishu in Somalia, where barriers of trees such as *Euphorbia biloculata* and *Commiphora* are also being tried.

3. Fixation of the dune itself

This can be done independently or in addition to (1.) and (2.) above, since revegetation with cactus and euphorbia is primarily for the purpose of sand fixation and has little economical value: tree planting is preferable; it consists of revegetation in three stages (Fig. 7):

a. In the first stage, living barriers are established to reduce the abrasive effect of moving sand on young trees;

b. Two or three years later pioneer trees should be planted either in the corridor or in the flat land between the linear dunes;

c. The third stage entails planting the lower third or half of the slope leaving the upper part of the slope bare for several years to allow wind to flatten the top of the dune by deflation. When the dune is flattened, then trees should be planted on the surface.

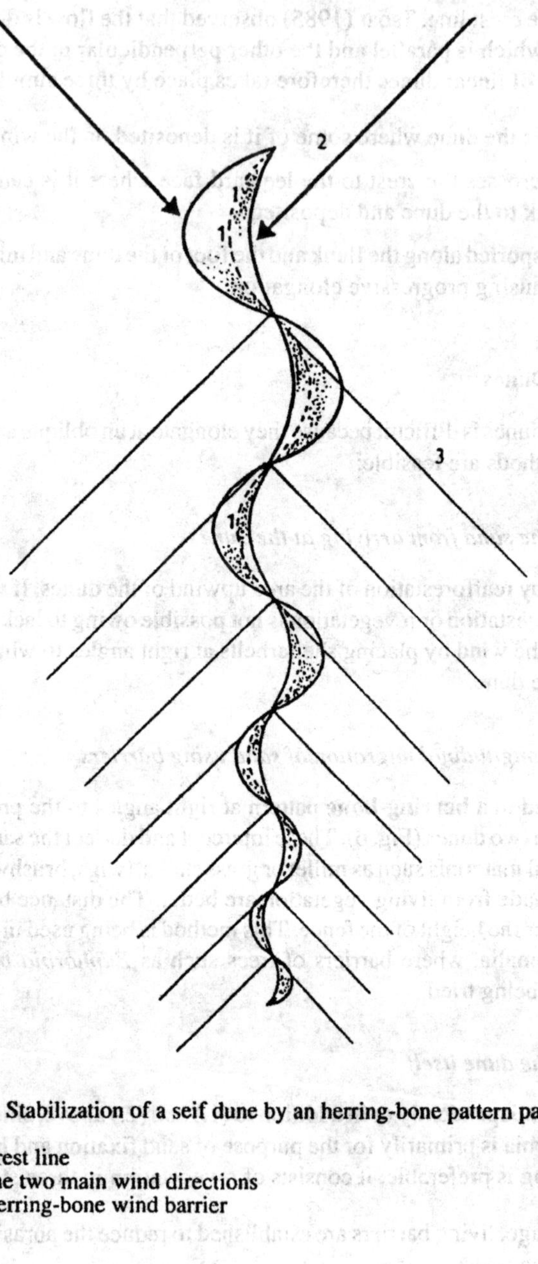

Figure 6. Stabilization of a seif dune by an herring-bone pattern palisade

1: crest line
2: the two main wind directions
3: herring-bone wind barrier

Stage 1:

Stage 2:

Stage 3:

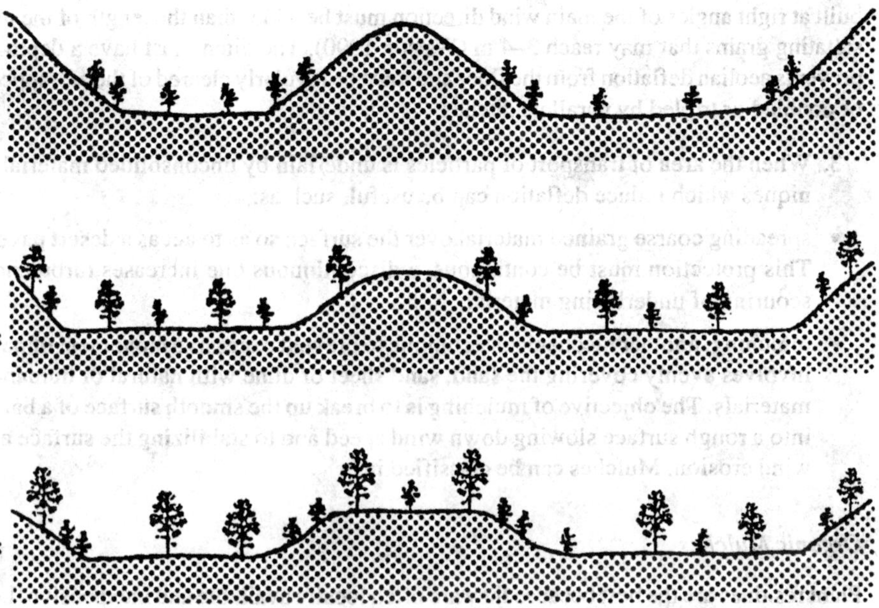

Figure 7. Dune revegetation.

BASIC PRINCIPLES OF STABILIZATION

In hyperarid and arid ecosystems, with an almost limitless supply of sand and with frequent and strong winds, two forces are active: mobile dunes and drifting sand. All the strategies to control aeolian damages linked with drifting sand, can be summarized by three actions: blocking, deviating, stabilizing which are often simultaneously used:

- **Blocking** the shifting sand at the source or at any step of transport by palisades or barriers.
- **Deviating** during the transport so as to avoid sand encroachment.
- **Stabilizing** or fixing the loose sand by mechanical, chemical or biological measures.

Moving sand in hyper-arid or arid ecosystems can be dealt with in five ways:

1. Blocking of the sand in the source area by wind barriers or artificial dune.
2. Along the area of transport of particles, the objective is to minimize the volume of particles transported, by using: ditches, palisades, or vegetal wind barriers, so as to encourage accumulation upwind of the area which requires protection. Nevertheless, to trap moving sand is a solution which needs continuous maintenance.

The excavation of deep ditches upwind of human settlements is based on sand entrapment and can bring temporary protection against encroaching sand. To be effective the ditches built at right angles of the main wind direction must be wider than the length of the jump of saltating grains that may reach 3–4 m (Watson 1990). The ditch must have a depth which prevents aeolian deflation from the floor and must be regularly cleared of the entrapped sand or doubled or tripled by parallele ditches.

3. When the area of transport of particles is underlain by unconsolided material, techniques which reduce deflation can be useful, such as:

- spreading coarse grained material over the surface so as to act as a desert pavement. This protection must be continuous, a discontinuous one increases turbulence and scouring of underlaying material;
- mulching, another mechanical technique to prevent excessive movement of particles involves evenly covering the sand, sand sheet or dune with natural or human-made materials. The objective of mulching is to break up the smooth surface of a bare field into a rough surface slowing down wind speed and to stabilizing the surface against wind erosion. Mulches can be classified in:

Organic Mulches

Plant residues are the best because they are often locally available and inexpensive. Leaves, twigs, branches or grasses can be laid on the soil surface or dug into the soil. Branches can also be bounded by wire, rope or strong grasses to construct small panels for covering the sand, sand sheet or dune.

Artificial Materials for Mulches

They include:

- sheets of plastic which can be used when the surface to fixed is small;
- acrylic fibre: 50 kg of fibre for each hectar covered;
- acrylic wire netting, (in rolls of 1.5 m wide and 5 m long) with three types mesh: 2.5 ×2.5; 3×3 and 5×5 mm;
- oil by-products such as asphalt or kerosene (paraffin);
- various chemicals.

Stabilization of Sandy Soil by Chemical Measures:

In extremely dry regions where planting vegetation is impossible and the transportation is very difficult, chemical methods are best for sand stabilization.

Kerosene (paraffin) is the most widely used oil by-product for stabilization. It has proven to be more effective than asphalt because it can penetrate the topsoil to a depth of 10–20 cm; its effects last for 3–4 years and still allows vegetation growth while stabilizing the sand surface.

Asphalt emulsion or polymers are more expensive than biological and mechanical methods. The cost for shifting sand stabilization with asphalt emulsion or polymer are 3 to 6 times higher than with straw checkerboard. So cheaper chemicals should be explored.

Experiments have proved that shifting-sand stabilization with chemicals is an efficient method which can be used to solve the problems of urgent dune stabilization such as along the railway lines or around industrial bases in sandy areas.

If the chemical method is combined with biological methods the sand-fixing layer formed by chemicals can be maintained in good condition until the vegetation is established and takes over the controlling function of wind erosion.

In areas where rainfall reaches 300 mm/year, in order to stabilize moving sand, the following steps should be taken: (1) a checkerboard pattern of low fences (25–30 cm high above the soil surface) is installed. Various materials can be used to construct fences with the checkerboard pattern, such as: bundles of millet stalks and other crop residues, plastic sheets and even cardboard; and (2) development of a natural or planted vegetation cover, the only sustainable solution to permanently immobilize any dune. This method was initiated by Chinese scientists in Shapotu, south of the Tengger desert.

4. All human settlements located in a sand current act as obstacles and therefore, reduce windspeed responsible of sand accumulation. If any solution to avoid this hindrance cannot be found, two palliatives can be proposed: 1. to deviating the sand flow by palisades or tree belts; 2. to enhancing sand movement by all kinds of aerodynamical facilities. It is recommended to minimize the obstacles or any features which may disrupt the airflow or give them an aerodynamical form to avoid sand accumulation. Linear installations (pipelines, roads, railways, canals) should run parallel to the direction of the dominant sand drift. If it is not possible to reduce the sand deposition on the windward slope of an embanked road or railway, the windward slope should be graded no steeper than 1:6 (22%) (Watson, in Goudie 1990). This value is similar to the windward slope of a barchan. Wind speed increases over an upward-sloping surface and therefore the aeolian sand-carrying capacity is augmented.

With steeper inclination the windward face of an embankment acts as a barrier and promotes sand accumulation. An angulous break on the top of the windward slope promotes a separation in the air flow. In the separation area the decrease of the speed results in sand deposition on the road surface. Watson (in Goudie 1990) writes "It should be stressed that the break in slope at the top of any embankment should be rounded not angular." For roads or railways located in cuttings the value of the slopes must also be taken into account because a downward incline decreases the wind speeds and the aeolian sand carrying capacity. "Slopes not exceeding 1:10 (13%) have been advocated" (Watson in Goudie 1990).

5. Physical methods to decrease sand movement are adopted in order to prevent the loss of topsoil. The main physical methods are: 1. wind ridging and deep ploughing, 2. minimum tillage, 3. scaring, 4. pitting, 5. covering sand surfaces with clay, 6. the creation of an artificial dune and the creation of a checkerboard structure of barriers (for more information, see Mainguet, 1992).

Other basic objectives of sand and sand sheet mechanical stabilization are to fix the sand for a period of time long enough to enable natural regeneration of the vegetation cover or planted vegetation to establish itself without watering. And in so doing to avoid the abrasive action of blowing sand upon young seedlings which have germinated during the rainy season, the burying of seedlings by drifting sand, deflation and exposure of the root system.

Mechanical stabilization is based on the use of non-living, organic or inorganic material to construct sand-binding barriers. This method is not ideal, however, because the effective life span of these structures is limited: three to five years for the checkerboards in the Tengger desert in China, for example. Sustainable stabilization of dunes can only be achieved by the development of a vegetation cover. Mechanical techniques used for sand stabilization are essentially designed for the protection of human settlements and villages, communication lines, transportation routes and precious agricultural land.

Sand blocking measures must mainly be considered as an assist for and stabilization measures. But there are certain conditions where they can be sufficient: rare and scattered barchans moving on a relatively flat surface and threatening human settlements or a road or railway. The distance available between the first barrier and the installation which must be protected is 150 m minimum. Nevertheless, if the source of sand is too rich, the method will not be efficient.

BIOLOGICAL STABILIZATION OF SAND

Preference must be given to **biological control measures** because they are at long-term the only sustainable ones. The mechanical methods are short-term measures when immediate action is required and are more difficult to maintain on the long term and generally, more costly. But biological methods become more difficult to use when the rainfall for an area is lower than 300 mm/year. In areas of recent revegetation, fencing to prevent the access of animals to these areas is very important. As well, fencing where sand has been reactivated, can also allow natural rehabilitation to occur. In areas of continuous sand deposition, for example, in the Chinese basin sand seas, the solutions are more difficult and more expensive because they are linked to mountainous geomorphology and a permanent sand supply.

Permanent stabilization of sand, sand sheets and dunes requires the development of a permanent vegetation cover. There are three possibilities for rehabilitation:

- natural regeneration of the vegetation cover;
- semi-natural regeneration of the vegetation cover; or
- creation of a vegetation cover.

To these methods can be added biological shelterbelts, windbreaks and wind barriers.

A biological wind barrier is a strip of trees, shrubs and/or grass which slows down the wind speed, reducing at the same time evaporation and overheating of the soil. When the strip of vegetation is larger than four rows of trees, the term **shelterbelt** is used. The first shelterbelts installed in 1955 in Cameroon, north of Mount Kapsiki used rows of *Cassia siamea.*

Windbreaks, on the other hand, refer to 2–4 rows of vegetation. In addition there are **mini-windbreaks**, which are wind barriers constructed of permanent grasses such as *Andropogon gayanus* which is often used for thatched roofs in the Sahel south of Sahara.

The physical shelter effects of a wind barrier depends on its height, it density or permeability, the texture of the vegetation used and the distance to the barrier itself. Fig. 8 shows

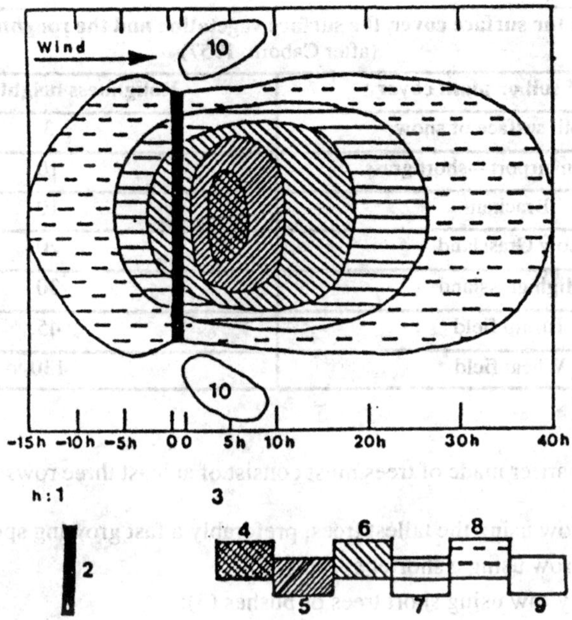

Figure 8. Distance from the wind barrier and reduction of the wind speed (after Gloyne, 1955).

1: height of the wind barrier
2: wind barrier
3: wind speed in percentage of the initial wind speed
4: < 20%
5: 20–40%
6: 40–60%
7: 60–80%
8: 80–100%
9: > or = 100%

the reduction of wind speed by a wind barrier in function of the distance from the barrier. The relationship between wind speed and height has been expressed by Geiger (1959):

$$V^2 = V^1.Z^{\circ}$$

where: V^2 = wind speed in m/sec.
Z = height of the measures in meters
V^1 = velocity to a height of 1 meter
° = 0.46 Karman constant

It is understood, therefore, that plant cover introduces a roughness or obstacle to the ground surface which decreases the wind speed to a certain height called the **roughness height** (Z°). Different types of vegetation have different values for roughness height.

Influence of the surface cover, the surface vegetation and the roughness height Z° (after Caborn, 1957)	
Kind of soil or plant cover	**Roughness height Z° (cm)**
Smooth surface of snow	3
Göttingen airport—short grass	10
Bracken	10
Low Grassland	20
High grassland	30
Turnip field	45
Wheat field	130

Ideally, a wind barrier made of trees must consist of at least three rows of plants (Fig. 9):

- the centre row using the tallest trees, preferably a fast growing species (1);
- the second row using a shorter species (2);
- the auxiliary row using short trees or bushes (3);
- wind direction (4).

Greenbelts, formed of several rows of trees or bushes, are a classic technique in China. The species are local ones selected based on their resistance, the speed of growth but also water

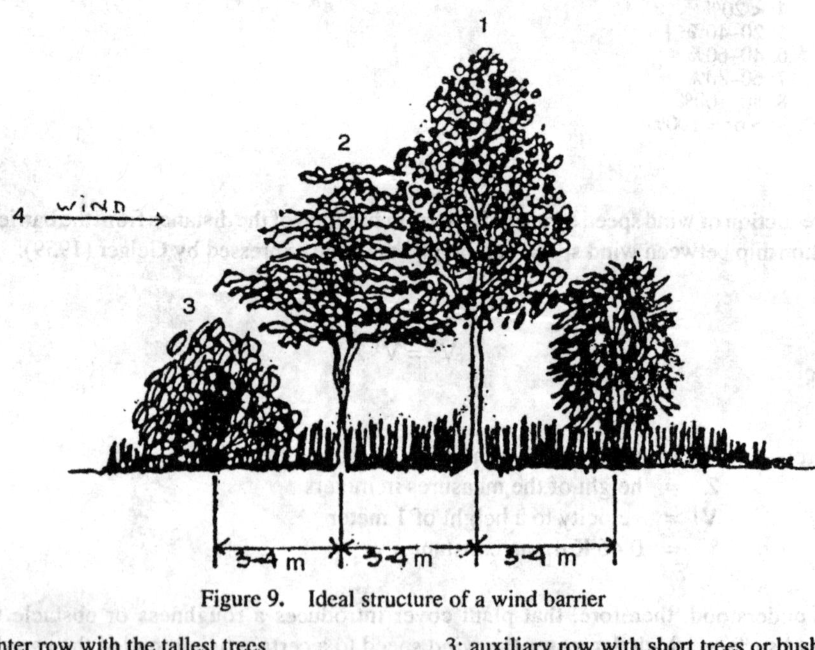

Figure 9. Ideal structure of a wind barrier

1: center row with the tallest trees
2: second row with shorter species
3: auxiliary row with short trees or bushes
4: wind direction

potential and depth of the underground water. In the Chinese oases, fruit trees are often planted: nut, apricot, mulberry tree.... In the oasis of Turfan (northeast of the Taklamakan), 16,000 ha of vineyard are surrounded and subdivided by greenbelts: 1130 ha were planted since 1964. Similar greenbelts are built in Hotian to avoid a bidirectional wind regime.

To obtain the maximum decrease in wind speed: the **ideal spacing** of barriers should be five times the height of the wind barrier. However, this optimum spacing may not be compatible with agriculture or land tenure in the area. According to wind speeds and topography, the **recommended spacing** would be between 5 to 25 times the height of the wind barriers. There is no general rule to predict the most efficient spacing. If the wind is very turbulent, for example, in areas with varying topography, in valleys swept by local winds or in corridor-cutting escarpments, the spacing must be reduced to 5 times the height of the barrier. The spacing of wind barriers depends on the topography of the windward slope where the wind streams are compressed and the wind speed is accelerated.

When the belts are aligned at 45 degrees to the eroding wind, and also on slopes exposed to the wind, the distance between the shelterbelts should be reduced. On the contrary, on the leeward slope, where the wind streams are expanded and the wind speed is slower, the density of the wind barrier can be decreased.

Intercropping is another biological technique to combat wind effects and improve production.

Micro-crop shelterbelt consisting, for example, of three rows of tall growing pearlmillet planted across the prevailing wind direction experienced by CAZRI in Radjastan in 1984 was found to be instrumental in increasing the water use efficiency and productivity of summer grown vegetables like lady's finger and cowpea. The pearlmillet crop for micro-crop shelterbelt have to be sown about a fortnight earlier to the actual sowing of vegetable crops for providing the shelter effect.

CONCLUSION

The success of any program of mobile sand control depends on the quality of the monitoring of the GWAS and all its parts. Air photographs, satellite images and field work supplemented with meteorological data are the tools to reach a good qualitative knowledge. Only these geomorphological tools can help engineers to chose the most efficient methods of sand control. These qualitative geomorphological investigations can be supported by quantitative assessments in laboratory or on field, for example wind tunnels or sand traps to measure the sand drift.

References

Anderson, R. S. and Haff, P. K., (1988). Wind modification and bed response during saltation of sand in air. *Acta Mech.* NATO Advanced Research Workshop on Sand, Dust and soil in their relation to aeolian and littoral processes, 14–18 May, Sandjberg, University of Aarhus, Denmark. In Acta Machanica/Supplementum 2, OE Barndorff-Nielsen and B. B. Willets, ed., p. 250, 83 figs.

Anderson, R. S., Sorensen, M., and Willetts, B. B. (1990). A review of recent progress in our understanding of aeolian sediment transport. Department of theoretical statistics, University of Aarhus, Research reports no. 213, p. 48.

Aufrère, L. (1931). Le cycle morphologique des dunes. Ann. Géogr., T 40, 226: 362–385.

Bagnold, R. A. (1941). The physics of blown sand and desert dunes. Methuen, London, p. 265.

Birot, P. (1981). Les processus d'érosion à la surface des continents. Masson, Paris, p. 579.

Breed, C. and Grow, T. (1979). Morphology and distribution of dunes and sand seas observed by remote sensing. In McKee ed.: A study of global sand seas. Washington, DC, USGPO: 253–305.

Bücher, A. (1986). Recherches sur les poussières minérales d'origine saharienne. Thèse de Doctorat d'Etat, Université de Reims, France, p. 290.

Caborn, J. M. (1957). The climate near the ground. In Geiger R. (ed.), Harvard Univ. Press, Cambridge, Massachusetts.

Chepil, W. S. (1945). Dynamics of Wind Erosion. Soil Science, 60: 305–320, 397–411, 475–780.

D'Almeida G. A. and Schutz, L. (1983). Number, mass and volume distribution of mineral aerosol and soils of the Sahara. Journal of Climate and Applied Meteorology, 22: 233–243.

El Baz, F. ed. (1984). Deserts and Arid Lands. The Hague, Martinus Nijhoff Publishers, p. 230.

Fryberger, S.G. (1979). Dunes forms and wind regime. In McKee ed. : A study of global sand seas. Washington, DC, USGPO.

Fryberger, S. G. , Al-Sari, A. M., Clisham, T. J., Risvi, S. A. R., and Al-Hinai, K. G. (1984). Wind sedimentation in the Jafurah sand sea, Saudi Arabia. Sedimentology, 31: 413–431.

Fryrear, D. W. (1981). Management of blank rows in dryland skip-row cotton. Transactions of the ASAE 24: 988–990.

Geiger, R. (1959). The climate near the ground. Harvard University Press, Cambridge, Massachusetts, USA.

Gloyne, R. M. (1955). Some effects of shelterbelts and windbreaks. Meteorol. Magazine no. 590.

Golinski, K. D. and Lindbeck, K. E. (1979). Woven plastic for coastal dune-forming fences. Journal of Soil Conservation Service of New South Wales, 35: 26–29.

Goudie, A. S. (ed.), (1990). Techniques for Desert Reclamation. Wiley and Sons, p. 271.

Lai, R. J. and Wu J. (1978). Wind erosion and deposition along a coastal sand dune. University of Delaware, Sea Grant College Programm DEL-SG-10-78, pp. 26.

Liu, Shu (1986). Basic ways of securing mobile sands in China. *Problems of Desert Development*, no. 3: 78–81.

Mainguet, M (1984). A classification of dunes based on aeolian dynamics and the sand budget. El-Baz ed: ***Deserts and Arid Lands***. The Hague, Martinus Nijhoff Publishers: 31–58.

Mainguet, M. (1991). Desertification. Natural Background and Human Mismanagement. Springer Verlag, Springer Series in Physical Environment, Vol. 9, p. 306.

Mainguet, M. (1992). Desertification through wind erosion and its control in Asia and the Pacific. Bangkok, ESCAP/UNDP, p. 135.

Mainguet, M., Canon, L., et Chemin M-C (1979). Le Sahara: géomorphologie et paléogéomorphologie éoliennes. In "The Sahara and the Nile. Quaternary environments and prehistoric occupation in northern Africa", M. A. J. Williams e. H. Faure éd, Balkéma, Rotterdam, Fasc. 2, pp. 17–35.

Mainguet, M., Cossus L. et Chapelle, A. -M. 1980 a. Utilisation des images Météosat pour préciser les trajectoires éoliennes au sol au Sahara et sur les marges sahéliennes: interprétation des documents Météosat du 28 mai 1978 au 9 février 1979. Soc. Fr. Photogrammétrie et de Télédétection, Bul. no. 78 (1980–2), pp1-12 + 3 images de Météosat.

Mainguet, M., Cossus L. et Chapelle, A. -M. (1980b). On the use of Meteosat Imagery for the determining of ground wind trajectories in the Sahara and regions bording the Sahel. Interpretation of Meteosat Images recorded between May 28, 1978 and February 9, 1979. Proc. 14th Intern. Symp. on Remote Sensing of Environment, 23–30 avril 1980, San-José, Costa Rica, ERIM, Ann. Arbor, USA, pp. 743–750.

Middelton, N. J. (1985). Effect of drought on dust production in the Sahel. Nature, 316, 6027: 431–434.

Pewe, T. L. (1981). Desert dust: an overview in desert dust: origin, characteristics and effects on man. The Geological Society of America, sp. pap., 186: 1–10. Fig. Deserts of the earth and major directions and distances of dust transport.

Prospero, J. M., Glaccum, R. A. and Nees, R. T. (1981). Atmospheric transport of soil dust from Africa to South America, Nature, 289: 570–572.

Tsoar, H. (1983). Dynamic processes acting on a longitudinal (seif) sand dune. Sedimentology, 30: 567–578.

Watson, A. (1991). The Control of Blowing Sand and Mobile Desert Dunes. In *Techniques for Desert Reclamation*, Goudie ed. pp. 35–85.

Willetts, B. B. and Rice, M. A. (1986). Collision in aeolian transport: the saltation/creep link. In Nickling W. G. ed: Aeolian Geomophology, Allen and Unwin, Boston: 1–17.

Zahirov, R. S. (1981). Prevention of sand drifts on railways, roads and irrigation systems. UNEP, Moscow, p. 182.

Zhu Zhenda (1990). Distribution of sand dunes in China, chapter I: 1–4, in The principles and measures of sand dunes stabilization in China, The Institute of Desert Research, Chinese Academy of Sciences, Lanzhou, p. 52.

POSTSCRIPT

When Iraq invaded Kuwait on 2 August 1990, no one could have foreseen what transpired during the six months that followed. Over half a million Iraqi troops were moved into Kuwait. An equivalent force from 30 countries of the "Coalition Forces", were amassed along the borders of Kuwait with Saudi Arabia.

As the liberation of Kuwait was initiated in mid January 1991, Iraq began to spill oil into the Gulf water from wells, pipelines, terminals and tankers. Shortly thereafter, Kuwait's oil wells were exploded as Iraqi forces began to retreat, and the whole world watched as a pall of black clouds began to emanate from the well fires.

A near hysteria of concern about potential effects on the global environment resulted in misleading propositions, and early predictions were made of a global environmental disaster with no scientific basis. For example, forecasts of a "nuclear winter" over Europe and the rest of the world were not based on knowledge of the wind directions in the Gulf region. They also did not take into account the fact that particulates from the burning wells were too heavy to rise to the stratosphere and spread around the globe. These particulates remained mostly below 8,000 feet and did not rise much above 22,000 feet. Thus, their effects remained in the region of the Arabian Peninsula and nearby areas of northwestern India and East Africa.

This volume presents scientific data as analyzed by an international team of scientists who are actively engaged in data gathering and analysis. The book is timely in view of heightened interest in environmental issues in the wake of the "Earth Summit" held in Rio de Janeiro in June 1992, and its emphasis on the need for assessments of environmental impacts based on sound scientific data.

The first chapter of this book introduces the reader to the geography of the Arabian Gulf region as an integral part of the Arabian Peninsula. It presents examples of pre- and postwar Landsat and SPOT images and Space Shuttle photographs to illustrate how spacecraft data can help in monitoring environmental change. It is followed by a chapter on the Gulf marine environment that was adversely affected by the oil spills. The slow, counterclockwise water currents of the Gulf moved the oil along the eastern coast of the Arabian Peninsula, endangering marine life, polluting hundreds of miles of coastline, and threatening water desalination plants. As lighter components of the oil evaporated from air exposure and the high temperature of the Gulf water, heavy components sank to the bottom to endanger sea grasses and shellfish breeding grounds.

The next three chapters are devoted to the wind regime, which was responsible for the distribution of particulates in the atmosphere. The first establishes the major circulation of air in the Gulf region with emphasis on high velocity winds of the winter and summer *Shamal*. The following chapter summarizes computer-based modeling of the wind flow in the Gulf region in order to make projections of the Kuwait oil fires pollution dispersion. Extensive meteorological models incorporating explicit terrain influences to the flow fields were employed through a six month period. Results show generally close agreement with satellite imagery of the smoke plumes. However, there are some examples of significant disagreement or failure of the meteorological models. These failures are most likely linked to missing or unavailable weather observations. The third discusses the dispersion pattern locally in Kuwait based on Gausian models, with emphasis on the sulphur dioxide component in the well fire plumes.

Chapter six discusses the potential impact of degradation of the environment on health, based on studies of exposure of Kuwaiti people through what they ate, drank, breathed and/or touched during the burning of the oil well fires. It emphasizes the growing concern that the war's consequences associated with exposure to emitted oil-related contaminants and the inherent uncertainty of any forecast, make the prediction and analysis of environmental impacts and risks a task of increasing importance. It also indicates that as public awareness of the importance of environmental health grows, the need exists for a careful definition of the association between the health of individuals, the pollution of the environment, and the extent of the effect on socio-economic circumstances. It further recommends the necessary steps for biomonitoring and human health environmental impact studies in the future.

The next chapter discusses the impact of the Gulf War on the desert ecosystem in Kuwait, with emphasis on environmentally protected areas. It presents the pre-war status of these, and documents the post-war changes, particularly on desert plants and animals. One interesting aspect of the discussed conditions is the unexpected rejuvenation of desert shrubs, including in areas badly affected by oil deposits.

Chapter eight emphasizes the effects of the war on the fragile desert surface through disturbance of the "desert pavement." The latter is usually a layer of pebbles, one grain thick, that protects the fine grained soil beneath from the action of the wind. Digging trenches, building of berms and other sand walls, and rapid movement of military vehicles removed the armor of pebbles and exposed the soil below to wind erosion. The result is an increase in the frequency of dust storms and the creation of sand dunes, some of which already block roads in northern Kuwait. The process appears to have been effective due to the fact that much of Kuwait's land surface is a deltaic deposit of a partially buried river that emanates from the Hijaz Mountains of western Saudi Arabia. This newly discovered relationship explains the prevalence of gravel deposits in northwestern Kuwait and the vast amount of fine grained soil beneath the desert pavement. Because of the tremendous disruption of the desert surface cover in Kuwait due to the Gulf War, the effects in the region may be long lasting.

The final chapter shows how wind action becomes a damaging process by soil erosion; one of the most efficient mechanisms of land degradation. It reviews the processes that act upon desert surfaces and the factors that control the amount of sand in transit. It also exemplifies the influences of topography on patterns of particle movement and sand accumulation, and concludes with a discussion of measures to control particle transport by the wind. These recommendations must be taken into account in remedial action of the effects of the Gulf War on the desert surface of Kuwait.

Although early predictions of a global environmental disaster in the wake of the Gulf War were proven unfounded, this book shows that much harm was done to the local environment. The environmental damage will have both short- and long-term consequences. The impacts on the local environment must be remedied to ameliorate the war's effects.

It is my opinion that the world community must learn from the sad experience of the environmental effects of the Gulf War. Perhaps one way of encouraging the collection of scientifically valid data is for the United Nations to sponsor the creation of an "International-al Center for Environmental Studies," to deal with major degradations that affect several countries or that have regional consequences, such as the Gulf War effects on the region. Efforts by individual countries may not provide the necessary broad view utilizing the same specifications and data collection methodologies in the whole region.

Furthermore, the need exists for a dedicated mission to monitor environmental change from space and to serve as an early warning system. Satellite images were crucial to the understanding and assessment of environmental damage due to the Gulf War. In order to prepare for the monitoring of events that cause damage to the environment, whether natural or man-made, I propose to launch "Envirosat," a satellite that would be dedicated to the monitoring environmental change in the future. Such an endeavor requires international cooperation in planning, execution and use of the acquired data, and may best be launched under the auspices of the United Nations.

Finally, the Gulf War has shown the need for a new "Environment Convention," along the lines of the "Geneva Convention," on the proliferation of nuclear weapons. The new convention would protect the Earth's environment from ruin in the course of military conflicts. In the wake of the Gulf War, the United Nations Compensation Commission was established to compensate Kuwait and other countries from the damage due to the war. Laws governing the compensation for environmental damage should be established and enforced to limit such potential in the future.

Farouk El-Baz

INDEX

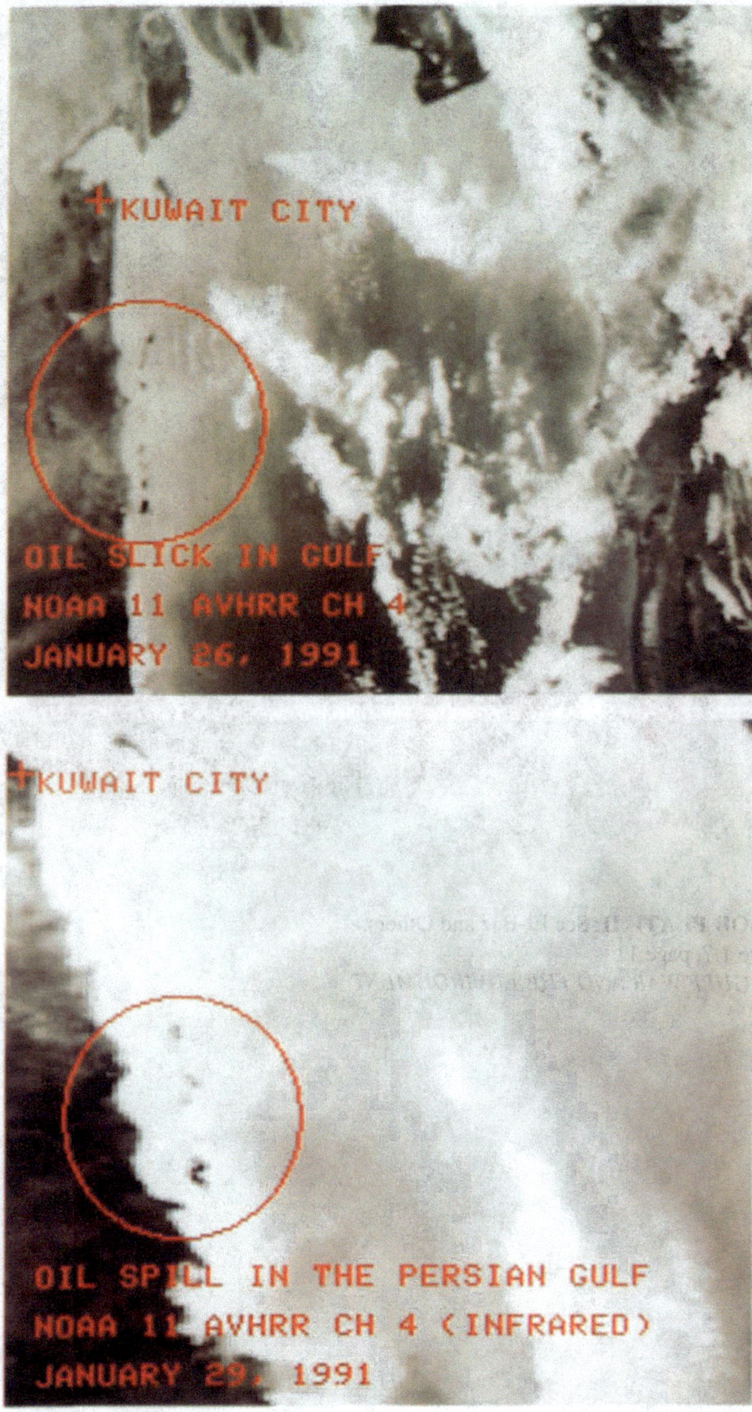

COLOR PLATE I. See El-Baz and Others,
Figure 1.3, page 6
THE GULF WAR AND THE ENVIRONMENT

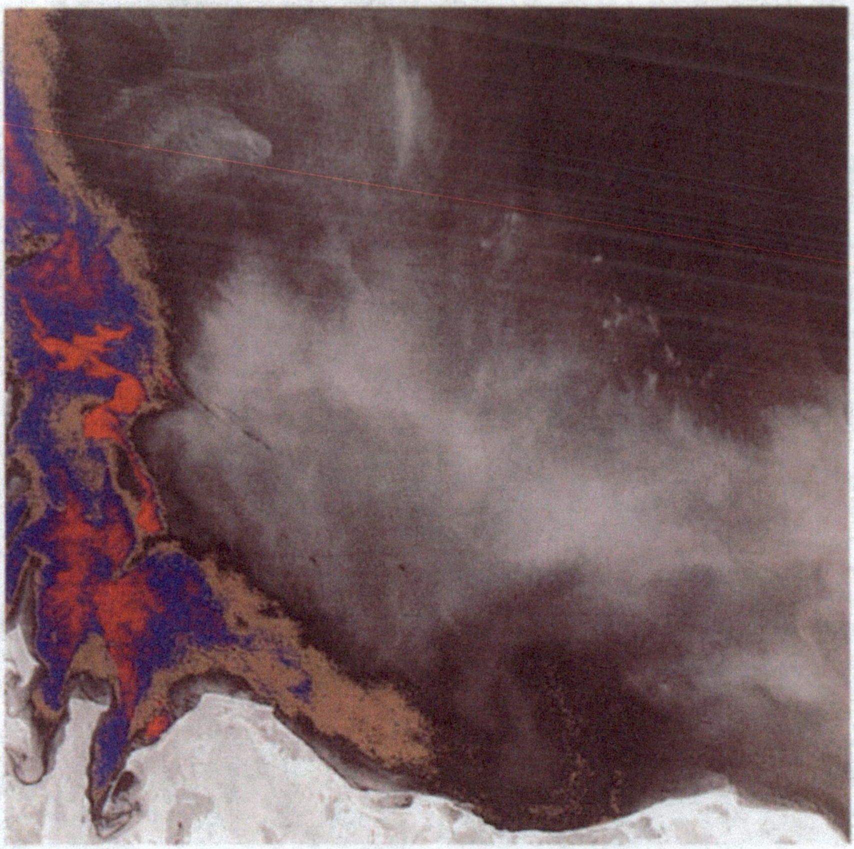

COLOR PLATE II. See El-Baz and Others, Figure 1.7, page 11
THE GULF WAR AND THE ENVIRONMENT

COLOR PLATE III. See El-Baz and Others, Figure 1.9, page 13
THE GULF WAR AND THE ENVIRONMENT

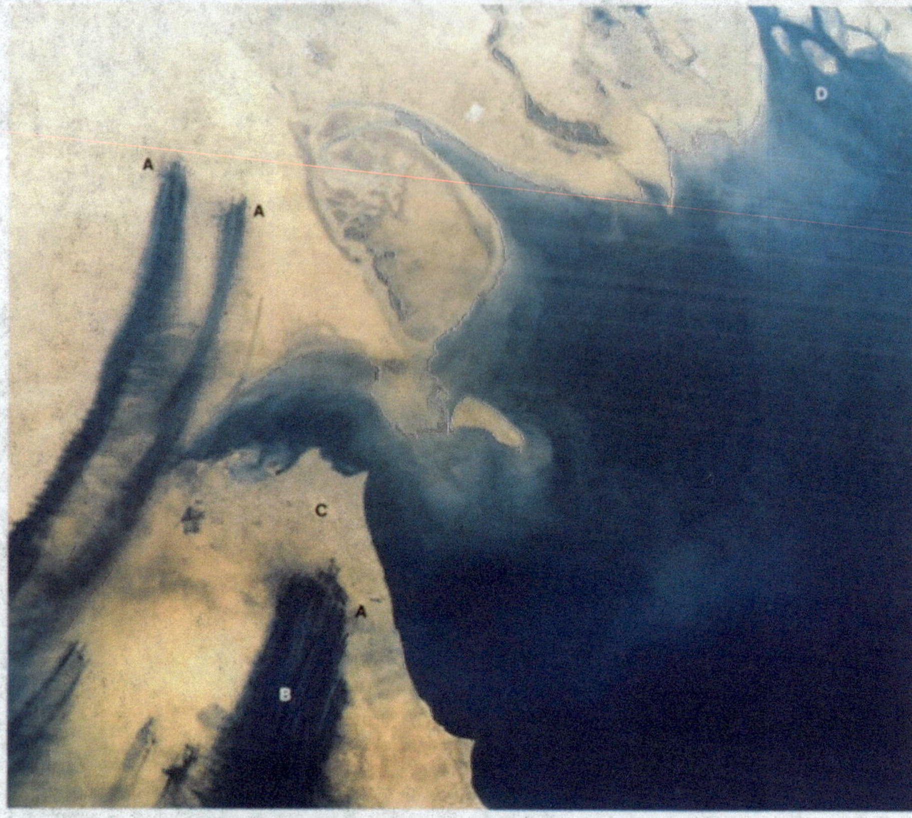

COLOR PLATE IV. See El-Baz and Others,
Figure 1.10, page 14
THE GULF WAR AND THE ENVIRONMENT

COLOR PLATE V. See El-Baz and Others,
Figure 7.10, page 127
THE GULF WAR AND THE ENVIRONMENT

COLOR PLATE VI. See El-Baz and Others, Figure 7.12, page 128
THE GULF WAR AND THE ENVIRONMENT

COLOR PLATE VII. See El-Baz,
Figure 8.2, page 135
THE GULF WAR AND THE ENVIRONMENT

COLOR PLATE VIII. See El-Baz,
Figure 8.8, page 145
THE GULF WAR AND THE ENVIRONMENT

9781138657304